Excel
Basic Skills

Problem Solving

5–6 Years
Ages 10-12

Get the Results You Want!

B Johns

PASCAL PRESS

Reprinted 2001, 2002, 2003, 2004, 2006, 2007, 2008, 2010, 2011 (twice), 2012, 2013, 2014, 2017 (twice), 2019, 2020 (twice), 2021, 2022, 2023, 2024

ISBN 978 1 74020 051 6

Pascal Press
PO Box 250
Glebe NSW 2037
(02) 9198 1748
www.pascalpress.com.au

Publisher: Vivienne Joannou
Edited by Beverley Weynton
Text design and typesetting by Tarnie Lowson
Cover by DiZign Pty Ltd
Printed by Vivar Printing/Green Giant Press

Contents

Year 5

Year 6

Introduction

This book provides an innovative approach to mathematical problem solving, and is designed to give Year 5 and Year 6 children the opportunity to practise problem-solving skills that have been taught at school. The book is divided into two sections: Year 5 and Year 6. Each section is divided into self-contained units, with easy-to-understand explanations and interesting activities. Students should be able to work through the book without supervision, and parents should be able to use it as a guide to explaining problem-solving concepts.

There are two units of work dedicated to each problem-solving strategy. Each unit consists of two pages: an example page, where the problems are worked through step-by-step and explanations and helpful hints are given; and a page of activities that require children to apply the particular strategy. All the activities involve problems that relate to children's everyday experiences, so they can easily visualise such situations. Both sections contain four revision tests to complete along the way and answers are supplied in a convenient lift-out section.

Finding a Pattern

In these examples we need to look for the difference as we move from one number to the next number in the series. The difference tells us about the pattern or sequence.

1. 3, 6, 9, 12, 15

> Work step-by-step to find the difference between the elements.

a) What is the difference between the first two numbers in this series? ____

b) The difference between element two and three is ____.

c) Does this difference apply to the others? Yes ☐ No ☐

d) The pattern is ________________________________.

e) The next three numbers in this series are: ____, ____, ____.

2. 2, 5, 9, 14, 20, ____, ____, ____

☐ ☐ ☐ ☐ ☐ ☐ ☐

a) The difference between 2 and 5 is ____. Record it in the box provided.

b) What's the difference between 5 and 9? ____. Enter that in the space provided.

c) Compare your answers so far: The difference is the same. ☐

The difference is changing. ☐

d) Subtract 9 from 14. The answer is 5. True ☐ False ☐

e) The next difference should be ____ if this pattern continues.

f) Subtract 14 from 20. Is your prediction correct / incorrect?

g) The next 3 differences should be ____, ____ and ____.

h) Therefore, the next 3 elements in the series should be ____, ____, ____.

3. 1, 2, 4, 8, 16, ____, ____, ____

1,	2,	4,	8,	16,
A	B	C	D	E

a) There are two ways of getting from A to B ________________________________

or ________________________________.

b) B to C can be done by ____________________ or ____________________.

c) To be a series, we can't have a changing variety of ways. Numbers are related.

Which way do you think is correct? ________________________________

d) Use your idea and check C to D. Correct ☐ Incorrect ☐

e) Does D to E fit your idea too? Yes ☐ No ☐

f) This is a ____________________ series.

g) Go back to the original series and enter the next 3 elements.

Activities

1. Find the next 3 elements in each of these series.

a) 30, 27, 24, 21, 18, ____, ____, ____ because the difference between each element is

______________________ .

b) 40, 33, 27, 22, 18 is the series. The next 3 elements are ____, ____, ____.

In this series, the differences were ______________________ .

c) (i) Here is the start of the series: 20, 15, 14, 9, 8

(ii) How was this series created? ______________________________

(iii) The next 2 elements are ____, ____

This problem involves finding the pattern in the series of numbers.

2. Captain Blood, the notorious pirate, ordered his crew to make a triangular stack of their store of cannonballs.
Here's how they made the stack.

top layer	2nd layer	3rd layer	4th layer	5th layer	6th layer	bottom layer
1	3	6	10			

a) The difference between the top and second layer is ____.

b) How many more (than the second layer) were needed for the 3rd layer? ____

c) What's the difference between the 3rd and 4th layers? ____

d) Look at the differences you have found. What's the pattern being formed? ____________

____

e) The next difference is ____ so the 5th layer should be ____ cannonballs.

f) There should be ____ cannonballs on the 6th layer because ______________________ .

g) The bottom layer will be made up of ____ cannonballs because ____ + ____ = ____.

h) We can check by showing this series as drawings. Finish the drawings below.

top layer	2nd layer	3rd layer	4th layer	5th layer	6th layer	bottom layer
●	● ● ●	● ● ● ● ● ●	● ● ● ● ● ● ● ● ● ●			

Finding a Pattern

Look at these triangles. There is a pattern to the markings.
Identify these changes to find the pattern. Predict the next element.

1.

A 

Triangle A has ____ vertical line and

____ shaded angles on the base.

B 

Triangle B has ____ vertical line,

____ horizontal line and

____ shaded angles on the base.

C 

Describe Triangle B ______________________________________

____

D 

Triangle D is the same as Triangle ____ with

____.

E 

If the pattern continues, then Triangle E should be the same as Triangle B. True ☐ False ☐

Therefore Triangle F should have ______________________________

____.

F 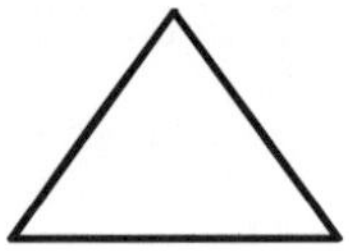

Complete the drawing.

Here's another pattern. These are circles. Look at each and identify the changes.
What's needed in the last circle?

2.

A 

Circle A has been divided into ____ equal sectors.

What's the difference between Circle A and B? ______________________

____

B 

In Circle C we have

☆ 3 equal sectors. True ☐ False ☐

C 

☆ one sector dotted. Yes ☐ No ☐

☆ one sector with __________ lines.

The difference between Circle C and D is that the last sector has

D 

____

E 

Circle E is the same as ____. This tells you the series will start ☐ end ☐ again.

If this series is to continue to the original pattern, then Circle F must have ____.

F 

You complete the last circle.

Activities

1.

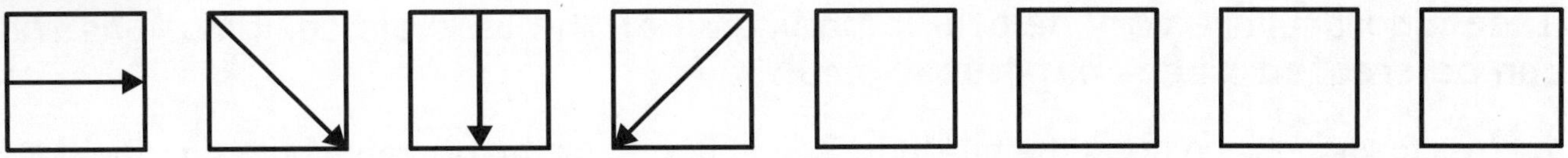

a) What's the same in each of these elements? ____________________

b) What's changing? ____________________

c) How are these changes taking place? ____________________

d) Finish off the drawings in this series.

2.

a) Every element in this series is based on a __________.

b) The number of __________ is changing.

c) The 5th circle should have ____ diameters. ____ diameters are needed for the 6th circle and ____ diameters must be drawn in the last circle.

3.

a) Every element in this series is a __________.

b) The only difference between the triangles is shape ☐ colour ☐ size ☐.

c) Measure the dimensions of the triangles and write them here. ____________________

d) Use your ruler and pencil to accurately draw triangles 5 and 6.

4. Gordon, the greengrocer, decided to make a display using a box of his best oranges. He decided that each layer would be in the shape of a square.

a) Complete the missing numbers and drawings in the table below.

layer	1	2	3	4	5	6	7
number	1	4		16			
shape					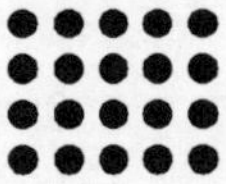		

b) Describe the pattern in this series ____________________

c) Exactly how many oranges were in the box he used? ____________________

Tree Diagrams

Tree diagrams are very helpful in finding all of the possible combinations that can be created. Check out this example.

1. In a townhouse complex the builder used either red, brown or cream bricks for the houses and tiled the roofs using green, orange or dark red tiles.
How many different colour combinations could be created?

☆ He was building ....

☆ We are considering..

☆ What are the colours? ....

townhouses
walls → red, brown, cream
roofs → green, orange, dark red

☆ Now let's take the options one at a time and list them.

red walls/green roof, red walls/orange roof, red walls/dark red roof
brown walls/green roof, brown walls/orange roof, brown walls/dark red roof
cream walls/green roof, cream walls/orange roof, cream walls/dark red roof

☆ Now let's count:

a) For red walls there are 3 possibilities.

b) For brown walls ____ combinations are possible.

c) ____ combinations can be created for cream walls.

d) This makes a total of ____ combinations.

2. If the builder decided to add one more colour (fawn) to the bricks, how many combinations are there now?

townhouses
walls → red, brown, cream, fawn
roofs → green, orange, dark red

a) List the combinations here.

	red walls/orange roof	
		brown walls/dark red roof
cream walls/green roof		
	fawn walls/orange roof	

b) There are now ____ combinations.

Note that by adding just one colour, more than one combination is added to the total.

c) If yet another brick colour was added, the total number of combinations would be ____.

Activities

1. Mum went to a sale and bought items which would mix and match.
She bought 3 blouses (red, white, blue), 2 skirts (white, black) and 2 pairs of shoes (navy, tan).
How many possible outfits can she create?

☆ We are investigating ....

☆ What are the items? ....

☆ The colours are....................................

☆ List the possible combinations.

Here's a start for you.

Do this in a logical way, taking one item at a time.

1 red blouse, black skirt, navy shoes

2 red blouse, black skirt, tan shoes

Now change the skirt.

3 red blouse, white skirt, navy shoes

4 red blouse, white skirt, tan shoes

a) Now change the colour of the blouse.

5 white blouse

6

7 ______________________ 8 ______________________

9 ______________________ 10 ______________________

11 ______________________ 12 ______________________

b) There are a total of _____ outfits she can create.

2. When she was ironing the white blouse, Mum scorched it.
Create a tree diagram to find how many outfits she has without the white blouse.

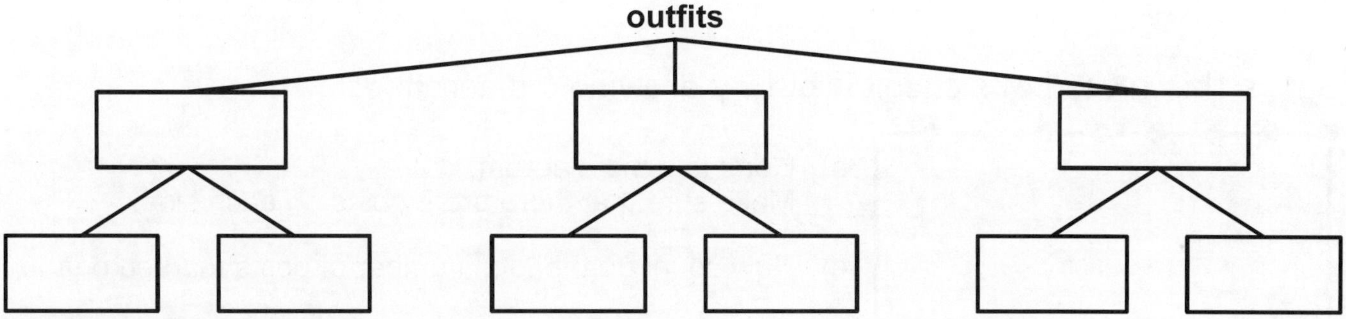

b) List the combinations. ______________________

____

____

____

c) Mum now has only _____ outfits.

Drawing a Picture

It is said that a picture is worth a thousand words. Often a sketch or diagram can help to solve a problem.

1. Six friends hired a boat to go fishing. It is important that the boat be evenly balanced with weight [side-to-side as well as back (stern) to front (bow)] so that it remains stable. Here are the masses of the people. A 42 kg, B 39 kg, C 44 kg, D 69 kg, E 35 kg, F 71 kg. Using the drawing below, arrange them so that the boat is evenly balanced.

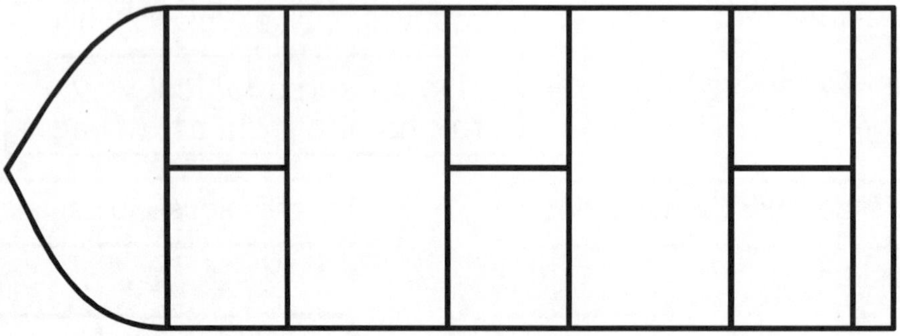

a) Find the total of all the masses. _______

b) This means that there should be as close as possible to ____ kg on each side of the boat.

c) Find two groups of 3 masses which will add to this number.

____ + ____ + ____ and ____ + ____ + ____ each equal ____ kg.

d) Now that you have the two groupings, pair them so that you have balance from bow to stern.
Enter their masses on the diagram and check out the balance.

2. Gina built a pen for her chickens. She framed a square shape and used 9 posts along each side. How many posts did she use altogether?

a) There are ____ posts on each side. There are ____ sides so that means there should be ____posts.

But is this true? Let's check it out by drawing a diagram.

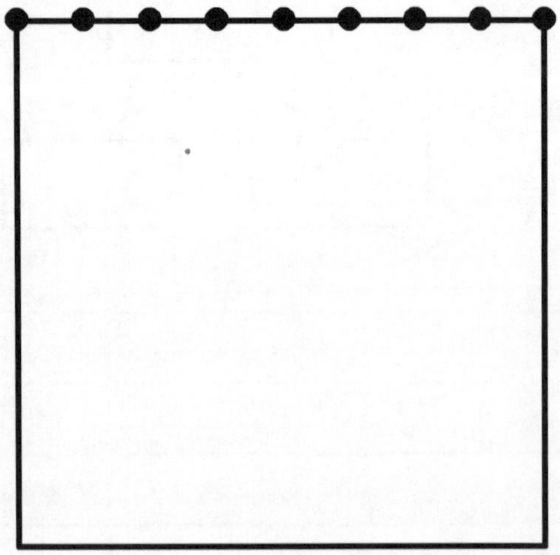

b) Complete the diagram.
Make sure that there are 9 posts on each side.

c) Now count up the total number of posts marked on the diagram.____

d) Does answer (a) agree with answer (c)? YES / NO.

Why? ______________________________

____

____

Year 5: Unit 4

Activities

1. I have 3 masses—9 kg, 5 kg and 3 kg—and a balance scale.

a) Arrange the masses so that the difference between the left-hand pan and the right-hand pan is only 1 kg.

b) How can I arrange the masses so that the difference is exactly 2 kg?

c) Show how I can weigh out 4 kg using only these masses.

d) Describe in words how you could use these three masses on the balance scale to measure exactly 7 kg? ______________________________

____

e) Draw pictures in the spaces below to show how to measure these amounts.

11 kg	17 kg	6 kg

2. Colin is a carpenter. He needs to measure out exactly 16 cm on a board but he's mislaid his tape. He does have 3 pieces of timber with their lengths marked on them: A 9 cm, B 11 cm, C 14 cm.

a) Draw a diagram which will give Colin a measurement of 16 cm.

b) Now describe this in words.

____

____

____

____

____

c) Which, in your opinion, is more easily understood – picture or words? ________________

Drawing a Picture

1. What length of wire is needed to go round a rectangular block of land (60 m long and the width is half the length) to form a three-strand wire fence?

a) What are the three important facts? ______________________________

____

b) Record this information on this sketch of the block of land.

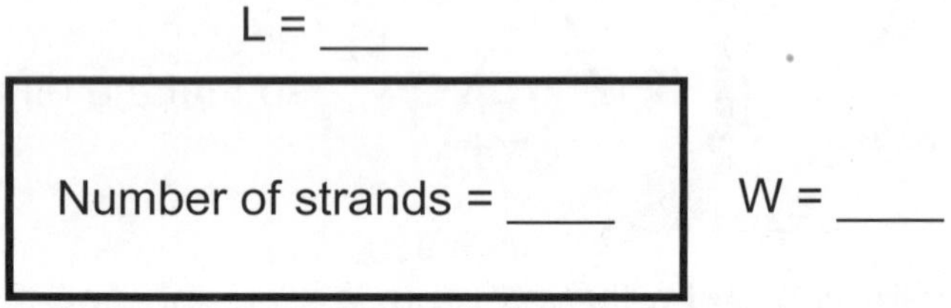

c) If the width is half the length, then the width is ____ metres.

d) Once round the block is ____m + ____m + ____m + ____m, which equals ____ metres.

e) The wire must go round this block ____ times.

f) Therefore, the total length of wire is ____m × ____ times and the answer is ____m.

2. Along the Great Western Highway are four towns: Ashton, Byfield, Canning and Dawson. Ashton to Byfield is 29 km. Ashton to Canning is 67 km, and Dawson is 132 km from Ashton. How far is it from Byfield to Dawson?

a) List the facts. ____

____

b) Complete this diagram and enter the information given.

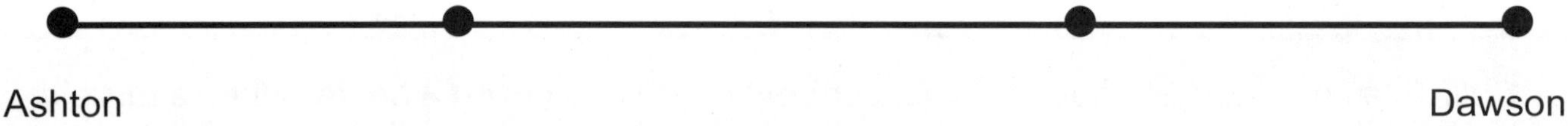

c) Now work out the distances between each of the towns. Ashton to Byfield ____km,

Byfield to Canning ____km and Canning to Dawson ____km.

d) What did the question ask? ______________________________________

e) How are you going to calculate this? _______________________________

____

f) The answer is ____km.

g) Ennavale is 38 km past Dawson. The total distance from Ashton to Ennavale is ____ km.

Explain how you found this. ______________________________________

____

Activities

Solve these problems using drawings/sketches.

Problem	Sketch/drawing	Answer
1. A mini-bus can seat 13 people (including the driver). How many trips (minimum number) will be needed to transport 37 people?		
2. A rain tank can hold 1000 litres. Since the last rain, we used 150 L, 200 L and 100 L. How much is left from a once-full tank?		
3. Work out the results of this race. Vern was 3 metres in front of June but 1 metre behind Carlos. The fourth placed competitor was Sharma.		
4. Eleven 'witches hats' were set out in a straight line. How far is it from the first to the last if each one is placed exactly 3 metres apart?		
5. Dad is building a 3-rail fence. He has 27 rails. This means that he will need ___ fence posts.		
6. Order these children from tallest to shortest. Maryann is 137 cm. Mike is 2 cm shorter. Molly is 5 cm taller than Mike. Masa is 4 cm above Maryann.		
7. Diamonds are very expensive. Gold is not as dear as diamonds but more valuable than silver. Platinum (white gold) is a little dearer than gold. Order these precious items from cheapest to dearest.		

Guessing and Checking

Some problems are best solved by making a guess, then checking that guess. This process continues until the correct answer is found. Try these puzzles.

1. Arrange the numbers 1 to 6 on this shape so that the total on each side is the same – 10.

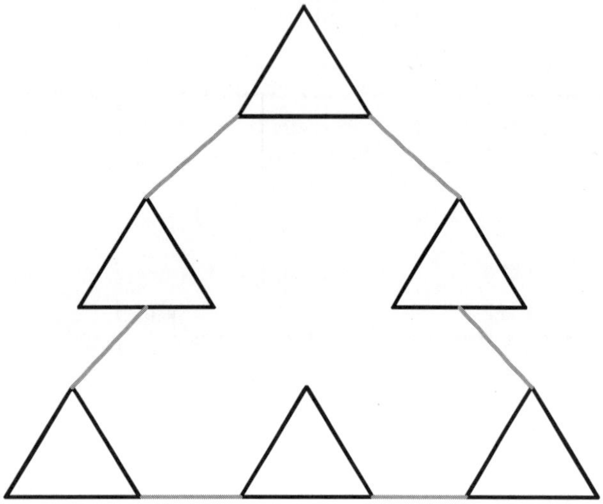

Every side must equal 10.
Use each number once only.

2. Arrange the same numbers 1 to 6 on this shape so that the totals on each of the three sides is twelve.

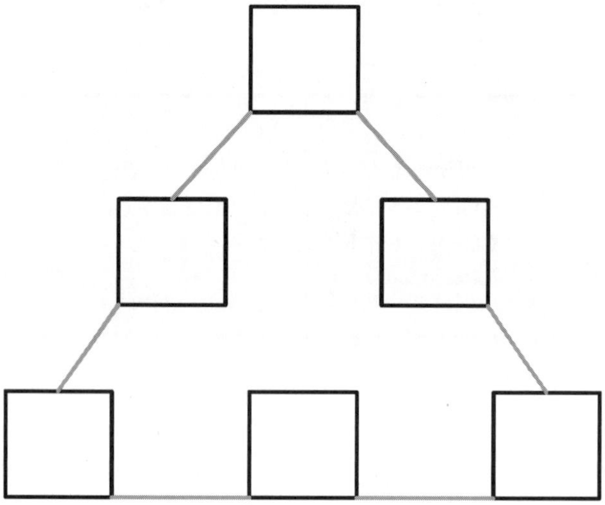

Use the ideas you gained from the first activity to help you solve this one.

3. Use the numbers 1 to 7 (each one only once) on this diagram so that each row of three circles has the same total.

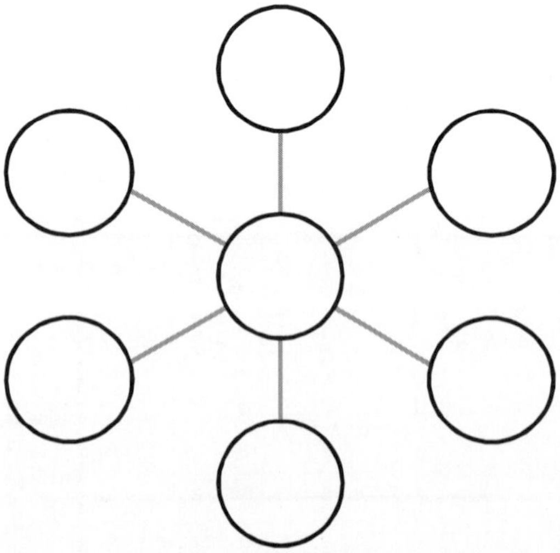

Check out other people's results. Are their totals the same as yours? Could there be more than one answer to this problem?

4. Use the numbers one to eleven once only in each of these circles so that the sum of each row of three circles is exactly the same.

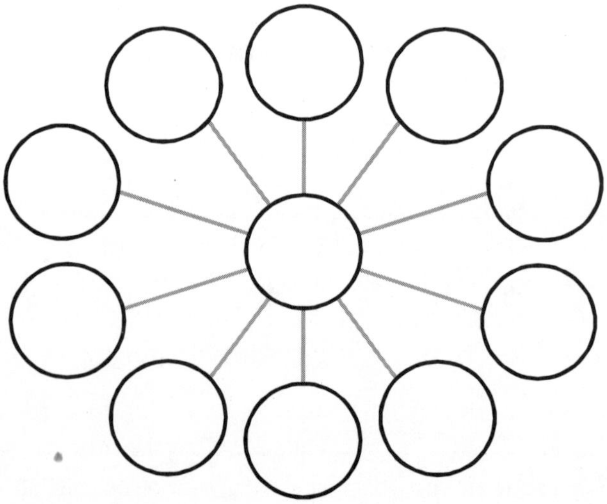

This looks very complicated because there are more circles, but the principle is exactly the same as the one before. You can solve this.

Activities

1. Make up an addition sum using digits 0 to 9 once only.
Place 3 digits on the top line, 3 digits on the second line and 4 digits in your answer.

Here are some boxes set out for you to use.

1st guess	2nd guess	3rd guess	4th guess
□□□	□□□	□□□	□□□
+ □□□	+ □□□	+ □□□	+ □□□
□□□□	□□□□	□□□□	□□□□

2. Here is a grid. How many squares can you count?

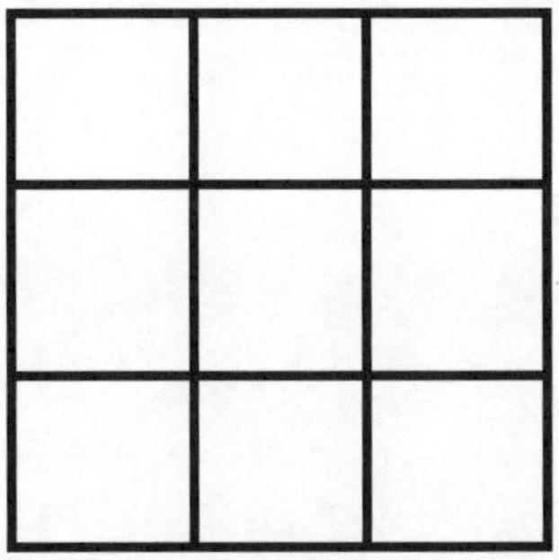

1st guess 9.

2nd guess ____.

3rd guess ____.

But what about the squares that have this shape?

Have you covered all of the possible shapes?

Check your results.

3. Solve these problems by using Guessing and Checking.

a) Think of a number. Add 6. Take 5. The answer is 4. What was the original number?

1st guess ____	check ✓ or ✗	2nd guess ____	check ✓ or ✗
3rd guess ____	check ✓ or ✗	4th guess ____	check ✓ or ✗

b) Think of a number. Double it, then take away 9. The answer is 3, so the original number must have been ____.

1st guess ____	check ✓ or ✗	2nd guess ____	check ✓ or ✗
3rd guess ____	check ✓ or ✗	4th guess ____	check ✓ or ✗

c) Find the number. I halved this number, then added 7. My answer is 12 so I started with ____.

1st guess ____	check ✓ or ✗	2nd guess ____	check ✓ or ✗
3rd guess ____	check ✓ or ✗	4th guess ____	check ✓ or ✗

Could these three 'Think of a number' problems be solved by using a different strategy? Working backwards, for example.

Guessing and Checking

1. Mum has 47 items of cutlery. There are two less forks than knives and 3 more spoons than forks. How many of each has she?

Here's a guess for you. | 10 knives, 8 forks, 11 spoons |

Check it out.

a) Are there two less forks than knives? YES / NO

b) Are there 3 more spoons than forks? YES / NO

c) When the 3 numbers are totalled, does it equal 47? YES / NO

d) Is this guess correct? YES / NO

e) If it isn't correct, what should you do? ______________________________

____

2nd guess: ____knives ____forks ____spoons

Check: Are all of the conditions set out in the problem met? YES / NO
If not, make a further guess.

3rd guess: ____knives ____ forks ____spoons

f) Answer: ____knives ____ forks ____spoons

Don't forget to check every guess to find out if you are right or wrong.

2. Callie has 6 coins. Their value is $5.50. How many of each coin has she?

		Check your guesses here
1st guess		
2nd guess		
3rd guess		

3. There are bicycles and tricycles. I counted 14 seats and 34 wheels.
How many of each were there?

	bicycles		tricycles		total	
	seats	wheels	seats	wheels	seats	wheels
1st guess						
2nd guess						
3rd guess						

Activities

1. Find these two numbers. When I add them the total is 47. If I take the smaller from the larger, the difference is 9. What are the two numbers?

Let's find the facts. ____ + ____ = 47 ____ - ____ = 9

a) Make your first guess. ____ + ____ = 47

b) Now see if the guess you made will fit the second fact ____ – ____ = 9

YES / NO. If no, what do you think you'll have to do? ______________________________

____

c) Make a second guess ____ + ____ = 47 ____ – ____ = 9

d) Is your second guess closer/further away? ______________

e) Use your second guess to make a third guess. Don't forget to check.

____ + ____ = 47 ____ – ____ = 9

The two numbers are ____ and ____.

2. From the set of numbers greater than 30 but less than 40, I chose three consecutive numbers. When I added them the total was 102. What are the three numbers?

a) List the numbers > 30 < 40.____

b) What are consecutive numbers?____

c) (i) Select a group of three consecutive numbers. ______________________________

(ii) Find the total. _______

(iii) Is the total equal to 102? YES / NO

d) If it isn't correct, make further guesses until your total is 102.

The numbers are ____, ____ and ____.

3. From the same set of numbers (>30 <40) I chose 3 numbers randomly. Their total was 107. What are the numbers?

a) The set of numbers is ____.

b) What does 'random' mean?____

1st guess ____, ____, ____	Check yes / no	2nd guess ____, ____, ____	Check right / wrong
3rd guess ____, ____, ____	correct / incorrect	c) The 3 numbers are ____, ____, ____.	

Year 5: Unit 8

Working Backwards

1. Double a number plus 8 equals 24. What was the number?

Let's start from the end and work backwards.

a) The answer given is ____.

b) To get this number, 8 was added, so the previous number was ____.

c) To get to that number, the original number was doubled. So to find the original number, you will have to halve it. The original number was ____.

d) Now check. Double your answer to (c). Now add 8, and your answer should be 24. This means your answer (c) is correct / incorrect.

2. I started with one number. I added 6 then doubled that.

My answer was 48 so this means that I started with ____.

Work backwards from what you know to find what you need to know.

a) The answer given is ____.

b) This answer is _______ what you had. This number is ____.

c) To find the original number, you'll have to add / subtract 6.

d) The original number was ____.

e) Describe how you checked your work. ____

3. $\square \times 7 + 5 = 47$

a) The answer is ____.

b) To find the second number you'll have to ____________ 5.

c) This number is a factor / multiple of 7.

d) Find the original number by _______________ by 7.

e) The number in $\square$ should be ____.

f) Check your work. Show your checking here. ____________________________________

____

4. $\square \times 3 - 8 = 22$

Describe how you solved this. ____

____

____

____

Activities

1. Jenny worked for 6 days earning the same amount each day.
When her earnings were added to the $32 she had in the bank her total balance was $98.
Each day Jenny earned $____.

This problem is the same as a number sentence but has words added.

a) Let's write it as a number sentence: ☐ × 6 + ___ = 98

b) To find her total earnings, you have to ______________________________________,

giving an amount of $_____.

c) If she earned this in 6 days, then she earned $____ per day.

d) Don't forget to check to prove you are right. Explain how you did this.

____

____

2. Tom is 3 cm taller than Brian.
Cassie is 5 cm shorter than Tom.
If Cassie is 136 cm tall, then Brian must be ____ metres tall.

a) How tall is Cassie? ________

b) Therefore, Tom must be _____ cm tall.

c) To find Brian's height you will have to ______________________________________ .

d) Go back to the question. Brian's height is required in centimetres / metres.

e) Give your answer correctly. Brian is ________.

f) Work forwards through the problem to check your working. I am correct / incorrect.

3. On Thursday Gold was selling at $390.
On Wednesday it was $15 less than on Thursday.
Tuesday's price was $12 more than Wednesday's price.
What was the value of gold on Tuesday?

a) Start with what you know. ____

b) Calculate the price on Wednesday. ________

c) Continue working backwards to find the price on Tuesday. ________

d) Check your working.

In each of these problems you started with what you were given and worked one step at a time, backwards, to find out what was required. Always check your work.

Working Backwards

Entering the data you are given on a flow chart and working backwards may help you see these problems more clearly.

1. I took 29 from a number then added 16.
My answer is 53, so the first number must have been ____.

Let's draw a flow chart for this.

( ) → (– 29) → (+ 16) → (= 53)

Now work backwards.

a) 53 ☐ 16 = ____

b) ____ ☐ 29 = ____

c) Enter your answer into the flow chart. Work forwards to prove your answer.

____

2. What number am I? I was doubled and had 10 taken away and ended up as 30.

a) Complete this flow chart.

( ) → ( ) → ( ) → ( )

b) The answer is ____.

c) Describe how you checked your work. ______________________________

____

3. Solve these flow charts by working backwards

a) ( ) → (- 6) → (+ 8) → (= 10)

Answer = ( ) ← ( ) ←

Check ______________________________

____

____

b) ( ) → (- 10 kg) → (× 2) → (= 10 kg)

Answer = ( ) ← ( ) ←

Check ______________________________

____

____

c) ( ) → (+ 6 m) → (÷ 2) → (= 9 m)

Answer = ( ) ← ( ) ←

I proved I was right by ______________________________

____

____

d) ( ) → (- $20) → (+ $36) → (= $36)

Answer = ( ) ← ( ) ←

The answer is correct because ______________________________

____

____

Activities

SEE, PLAN, DO and CHECK is the best way to approach all problems.
Use this method to solve this problem.

1. Over a 3-day period the Carlyle family drove from Broken Hill to Sydney. On the first day they drove from Broken Hill to Ivanhoe. Their next leg took them from Ivanhoe to Forbes – 126 km more than the first day. On the third day they completed their journey (Forbes to Sydney), 53 km less than the day before but a distance of 392 km.
Just how far did they travel from Broken Hill to Sydney?

A SEE

a) Read the problem to get an idea of what it is about.

b) Read the problem to make sure you understand it.

c) Go through the problem once again and highlight the QUESTION.

B PLAN

d) List the FACTS that you are going to use. ______________________________

____

e) How are you going to solve this problem?

____

C DO

f) How far did they travel on the third day? ________

g) This distance was ____ km more / less than the day before.

h) From this fact, calculate the distance they travelled on the second day. ____ ____ = ____

i) The difference between the distance travelled on the first and second days was ____.

j) If they travelled 445 km on the second day, then they travelled ____ on the first day.

k) Let's gather up the three distances. Day 1: ____ km, Day 2: ____ km and Day 3: ____ km.

l) What operation will we need to find the total? ____________________

m) Find the total: ____ km.

D CHECK

n) Does my answer seem correct? YES / NO

o) Is my working correct? Enter your answer into your calculator. Subtract the Day 1 distance. Now take off the distance for Day 2. Finally, take away the distance for Day 3.

If you've added correctly, then the answer on your calculator should read ________.

Acting It Out

This strategy is like drawing a picture but instead of a drawing you use real things or people. It makes the problem easy to SEE.

Shaking hands

1. Bill met Bob and they shook hands. How many handshakes?

a) Bill shook hands with Bob. Number of handshakes for Bill = ____.

b) Bob has already shaken hands, so his number of handshakes = ____.

c) The total number is ____.

2. If 3 people (Sue, Sandy and Sam) met and shook hands, there would be ____ handshakes.

a) Sue shook hands with _______ and _______ = _______ handshakes.

b) Sandy shook hands with _______ and _______ = _______ handshakes.

c) Sam has already shaken hands with the other two = ____ handshakes.

d) What is the total number of handshakes? ____

3. Let's increase the number of people to 5.

Gather a group of five people and follow the plan.

a) The first person shakes hands with ____ people.

b) The second makes ____ handshakes.

c) ____ handshakes are made by the third person.

d) Person 4 shakes hands with ____ people.

e) The last person makes ____ handshakes because ______________________________

____

f) List the numbers here ____, ____, ____, ____, ____

g) Find the total. ____

4. Nine people met. Each shook hands with every other person. In all, ____ greetings were made.

a) Person A: ____ handshakes.

b) How many people did B greet? ____

c) C made ____ handshakes.

d) ____ handshakes were made by D.

e) E shook hands with ____ other people.

f) F had to greet ____ people.

g) G greeted ____ different people.

h) How many people were left for H to meet? ____

i) I was the last with ___ left to meet.

j) List the numbers of handshakes.______________________________

k) Find the total. ____

Activities

1. The Wilson family comprises Mum, Dad, two boys (Jim and Jake) and three girls (Sue, Andrea and Paula). At Christmas time, each person gives everyone in the family a gift. How many gifts are exchanged?

Use counters, buttons, blocks... to represent the presents.

a) Mum gives presents to ______, ______, ______, ______, ______, ______ a total of ______.

b) Dad buys presents for ____ people.

c) How many presents will Jim have to buy?____

d) Jack gives ____ presents.

e) Each of the girls will need ____ presents each.

f) The total number of presents for the girls is ____ × ____ = ____.

g) Now gather up all of the counters (presents) and find the total. ____

2. Here is a die (the singular of dice). After it has been rolled over five times in a clockwise direction with the three continuing to face you, what will the side facing you appear as?

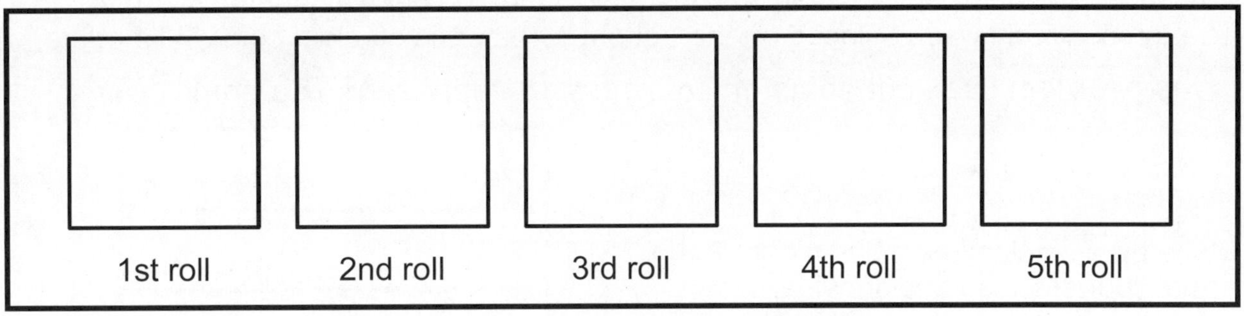

Use a real die. Follow the directions and draw the end result.

3. Jenny paid $1.60 for a brooch. She sold it to Carla for $2.00.
Carla decided she didn't like it so Jenny bought it back for $1.80.
Christine then purchased it for $2.20. How much money did Jenny make on the deals?

Use real money. Conduct the transactions and work out the profit/loss Jenny made.

a) Write down the steps that you took to solve this problem.

____

____

____

____

b) Jenny made a profit/loss (circle one) of $_______.

Acting It Out

1. In a line, there are two boys standing in front of 3 other boys. There are four boys behind one boy and one boy in the middle with two boys in front and two boys behind. What is the least number of boys in this line?

Use 8 different coloured counters for the boys. Try different numbers until you find the least number which will satisfy the conditions given in the problem.

a) Show a drawing of the number of boys in the line.

b) Prove that you are right by checking back to the problem.

Condition 1 two boys in front of 3 other boys YES / NO

2

3

2. Talia had 5 coins ($2, $1, 50 c, 20 c and 10 c). She stacked them up so that the $2 was above the $1 coin. The 50 c coin was just below the 20c but was above the 10 c. The ten cent coin wasn't placed on the bottom. Create the stack of coins Talia made.

For this problem use real coins or counters to represent the coins.

a) Make up a stack of the coins. Label your stack on this drawing.

b) Check the conditions from the problem.

(i) Is the $2 above the $1? YES / NO

(ii) Where is the 50 c? ____________________

(iii) Is the 20 c just below the 50 c? YES / NO

(iv) The ten cent is ____________________ .

(v) Is the ten cent on the bottom? YES / NO

c) What must you do if all of the conditions are not met? ____________________

____

d) List the coins in order from top to bottom. ____________________

Activities

A group of counters will help you solve this problem.

1. Nine people (Arthur, Betty, Charles, Don, Ethel, Fred, George, Helen and Iris) were in alphabetical order in a line. Starting with Arthur, they counted aloud in ones. Those who said an even number dropped out. Those left in the line continued counting. Again the even numbers dropped out. Those left continued until only one person remained. Who was that?

a) Write the nine names in order in the table below.

b) Write their numbers below each one.

c) Who were left? Write them.

d) Continue on numbering.

e) Who still remains in line?

____

f) Their numbers are ___________________.

g) Who is the only person left after crossing out the even number?_________________

2. If the same group of people repeated this activity but this time the odd numbers dropped out, who would be the last person?

a) Whom do you think it will be?___________________

b) Show how you worked this out.

1st counting

2nd stage

final step

c) Who's left? _________________

Creating a Table

1. Show all of the possible combinations using 20 c, 10 c and 5 c coins that will equal 60 cents.

I've drawn up the table and entered some data. You fill in the missing elements. Enter the data in a logical sequence. Note how I began.

20 cent coins	3	2	2	2	1	1			1						0	
10 cent coins	0	2	1	0	4	3										
5 cent coins	0	0	2	4	0	2										12

2. Marla has six counters (red, white and blue) with at least one of any colour. Create a table to show all the possible combinations of numbers of counters.

a) How many rows will you need? ____ Why? ______________________________

____

b) How many columns will you need? ____ (estimate)

c) Complete this table. Add extra columns if needed.

red							

Enter your data in a logical way. Start with 4 red 1 white 1 blue, then 3 red 2 white 1 blue and continue until 1 red 4 white 1 blue. Now increase the blue and decrease the white.

d) How many possible combinations were there? ____

e) Let's check by Acting It Out. Use counters to model the combinations.

Using this method there were ____ combinations.

f) Compare your results using the two strategies. SAME / DIFFERENT.

3. If the conditions of the previous problem were changed – you must have at least 2 of every colour – would this affect the results? Show the results on this table.

red							
white							
blue							

Activities

1. There were ducks and dogs in the backyard. The total number of animals was 10. The total number of feet was 26. How many of each animal were in the yard?

Use this table to enter numbers.

a)

Ducks	number	1	2	3	4	5	6	7	8	9
	feet	2								18
Dogs	number	9								1
	feet	36								4
Totals	number	10								10
	feet	38								22

b) Another strategy you could use to solve this is Guessing and Checking. Guess a number of ducks. This gives you the number of dogs, then you work out the total number of feet. If you weren't right, you'd guess again and check again.

c) Cover up the table and solve this by Guessing and Checking: 'Ducks and dogs – 10 animals – 30 feet'.

Answer is ____ ducks and ____ dogs.
Uncover the table and compare your Guessing and Checking with the table.

2. Create a table of all the possible ways $1 can be made using 50 c, 20 c and 10 c coins. Answer the questions by using your table.

a) 7 coins which total $1 are

______________________________________ .

b) I received 5 coins equal to $1 in value.

What were these coins?

____

c) The least number of coins to equal $1

is ______________________________________ .

d) The most coins received would be ________ .

e) Is there a way of receiving only 3 coins

to equal $1?______________________________

50 c	20 c	10 c	Total no. of coins

Creating a Table

1. Here's a list of some items sold at Burger Jack's. Coffee $1.50, Big Burger $2.95, Cheeseburger $2.30, Chips $1.50, Soft drink $1.20, Wedges $1.80, Chicken Burger $2.80 and Milkshake $1.90.

Let's take this disorganised list and put it into the form of a table.

Tables are much easier to read than random lists like that.

a)

Foods	Drinks

Use your table to work out these.

b) The Hudson family bought
2 Big Burgers ______
2 Cheeseburgers ______
1 Coffee ______
3 Soft drinks ______
Total $______
Change from $20 $______

c) Dallas spent exactly $6.50 on 3 items. Which items did he purchase?

____

d) Katarina has $10 to spend for herself and her little brother. He wants a Cheeseburger, Wedges and a Soft drink. She would like Chips, a Milkshake and a Chicken Burger.

(i) Has she enough money? YES / NO

(ii) If 'yes' how much change should she receive? ______

(iii) If 'no' by how much is she short? ______

e) On one day these items were sold. Calculate how much money was paid to Burger Jack's. Use the table and a calculator to work it out.

Item	Number Sold	Value of Item	Total Value
Big Burger	26		
Cheeseburger	10		
Chicken Burger	18		
Chips	21		
Wedges	6		
Coffee	5		
Soft drink	19		
Milkshake	13		
		TOTAL SALES	

Activities

1. Margaret's boss offered her the choice of two salary packages for the next 12 months:

A $1 the first month then doubling every month OR

B $300 a month, every month.

a) Which package do you think Margaret should take? _____

Before she makes a decision let's work it out.

PACKAGE A

Month	J	F	M	A	M	J	J	A	S	O	N	D
Earnings	$1	$2	$4	$8								
Total Earnings	$1	$3	$7									

b) Complete the missing elements in this table.

c) Her total earnings for the year will be $________.

PACKAGE B

d) Calculate her income - $300/month for 12 months = $________.

e) Changed your mind? YES / NO Why? ____

____

2. A laser printer can produce 9 sheets of word processing per minute.

a) Create a table here showing the production rate.

b) Use your table to answer these questions.

(i) Graham needs exactly 6 minutes to print his document. That document has ____ pages.

(ii) Trish has a 27-page report to print. This will take ____ minutes.

(iii) A fifty-page file will take between ____ and ____ minutes.

(iv) Mike spent from 10.05 a.m. to 10.11 a.m. printing his work.

His report contained ____ pages.

(v) A ream of paper is 500 sheets. Could this printer consume this amount in an hour?

YES / NO. Explain how you worked this out from your table.________________

____

Creating a List

A list is much like a table. It is an organised way to present data.

1. Ten people met for the first time. Each person shook hands with every other person. How many handshakes took place?

The list has been started. You follow the logical pattern.

a) Person A shook hands with B, C, D, E, F, G, H, I and J = 9 handshakes

B C, D, E, F, G, H, I and J =

C

D

E

F

G

H

I

J

b) Now find the total number. ____

c) Why didn't B shake hands with A? ______________________________

____

2. How many possible different motor cars can be created if a purchaser has the option of manual or automatic, V6 or V8 and 3 colours: red, black or white?

a) How are you going to solve this?______________________________

b) Create your lists here.

manual, V6, red

automatic, V6, red

c) There are ____ possible combinations.

d) List the combinations. ______________________________

____

____

e) Does this answer agree with your previous answer? YES / NO.

Activities

1. Here are the sales figures for fishing rods at Tom's Tackle Shop: January 96, July 32, March 24, May 65, February 37, April 58, June 63.

a) Re-organise this list into a logical order and answer these questions.

b) Best sales were in ______________.

c) In June Tom sold ____ more / less than in the following month.

d) In the first 3 months, he sold ____ more / less than in the last 3 months.

e) His total sales for the 7 months was ____.

f) Arrange the months in order from least sales to most sales.

____

2. These are the daily takings from Bob's Bakery last week: Tuesday $265, Monday $198, Friday $372, Thursday $102, Wednesday $248.

a) Create 3 separate lists:

in order of days	in order best sales to worst sales	least sales to best sales

You can see that there are several logical ways of presenting a list.
Use the lists to answer these questions.

b) Bob's best day was ______________.

c) The day his takings were closest to $200 was ______________.

d) How much did he take for the whole week? _____

e) On Tuesday his takings (compared to Friday) were down by $_____.

f) If he wanted to open only 4 days a week, on which day, in your opinion, should he close?

______________ Why? ______________________________

____

Remember in problem solving - SEE, PLAN, DO, CHECK.

Year 5:

Test 1

You can use either Finding a Pattern or Tree Diagrams to solve these.

1. Complete these.

a) 33, 38, 43, 48, ____, ____, ____

b) ●/●●, ●●/●●●, ●●●/●●●●, ●●●●/●●●●●, ____, ____

c) triangle, __________, __________, __________, heptagon, octagon

d) 2002, 1902, 1802, 1702, _______, _______, _______

2. It is said that one year in a dog's life is equivalent to seven years in a human being's life. Complete this table.

a)

dog years	1		3			6		
human years equivalent		14		28		42		

b) Mike (35 years old) reckons he and his dog are the same age. Just how old is his dog? ____

c) Our family dog is old by dog standards. She's almost 10 years old. This makes her almost equivalent to a ____-year-old person.

3. Work out all of the sandwich combinations which can be made using white, brown or wholemeal bread with the choice of any one of these four fillings – ham, salad, cheese, vegemite.

a) Show the diagram here.

b) List all of the combinations.

c) If another type of bread, e.g. pita, is added to the list, then there will be ____ combinations.

d) Make a list of these extra combinations. ______________________________

____

Year 5:

Test 2

Choose from either Guessing and Checking or Drawing a Picture to solve these.

1. Dad is putting up a fence along a 30-metre-long boundary. He'll put in posts every 2 metres, then fix rails to the posts. Work out how many posts he'll need, and the total length of rail timber.

a) Number of posts = ____ Why? ____

____

b) Length of rail = ____ Why? ____

____

2. Either prove or disprove this statement:
Jack set up 'witches hats' in the form of an equilateral triangle. Solaya counted 8 on each side. Paul said that means he used 24 hats to make the triangle.

a) Diagram

b) The statement is TRUE / FALSE because ____________________

____

3. a) I told Jo a number. To that number she added 7. Mike doubled Jo's number; the answer was 24. What was my number? ____

b) On sale were 74 items of crockery (cups, saucers and plates). There were 6 less cups than saucers and four more saucers than plates. Jenna started counting only the cups, then said she could work out the other quantities knowing just that.

cups						
saucers						
plates						
total						

Use this grid to help you. Answer: ____ cups ____ saucers ____ plates.

c) At the bikie convention there were motorbikes and motortrikes (the three-wheeler version) everywhere. I counted 74 seats. My sister counted 159 wheels. How many of each bike were at the convention?

bikes	number						
	wheels						
trikes	number						
	wheels						
total	number						
	wheels						

Year 5:

Test 3

Use either Working Backwards or Acting It Out here.

1. Jane had a stack of counters. Create her stack given these facts. Yellow and Black are above White. Red is on top of Blue. Yellow is just below Blue.

The correct order is (top)____ (bottom).

2. I had some money. I spent $35 on videos. Dad paid me $25 to help him put down a heap of bark in the garden. When I checked my purse I had $50.

How much did I have before my spending? _____

3. Show either in word form or diagram form how you would solve this number sentence:

$\square \times 5 - 8 = 27$.

4. It was the custom among some of the Indian tribes of North America to rub noses when they met a friend.

a) If there was a gathering of 8 friends, how many times would this greeting be performed if everyone greeted every other person? ____

b) At another tribal gathering, this greeting was repeated 66 times.

How many people were at this gathering?____

5. In one group there were two mothers, one grandmother, two daughters and one granddaughter. What is the least number of people who made up that group? ____

6.

If the box is rolled over 6 times in an anti-clockwise direction, with the name kept facing you, how will the front of the box then appear?

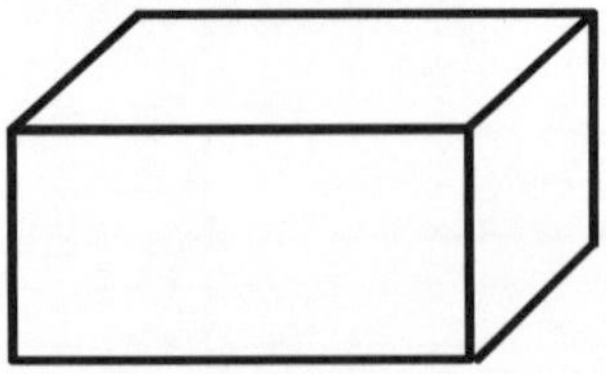

Show how the word will be written on this box.

7. a) $(\square - 5) \times 8 = 56$ $\square$ = ____

b) $56 - (\square \times 8) = 8$ $\square$ = ____

c) $(56 - \square) \times 8 = 8$ $\square$ = ____

The use of brackets can make a difference to your answers.

Answers

Page 2
1.a) 3 b) 3 c) Yes d) add 3 to previous element e) 18, 21, 24
2.a) 3 b) 4 c) changing d) True e) 6 f) correct g) 7, 8, 9
h) 27, 35, 44
3.a) + 1 or x 2 b) + 2 or x 2 c) x 2 d) Correct e) Yes
f) doubling or x 2 g) 32, 64, 128

Page 3
1.a) 15, 12, 9 subtract 3 b) 15, 13, 12 differences 7, 6, 5, 4, 3, 2, 1 c) (i) subtract 5, subtract 1 (ii) 3, 2
2.a) 2 b) 3 c) 4 d) adding one more than the previous difference e) 5, 15 f) 21, the difference is + 6
g) 28, 21 + 7 = 28 h) check drawings

Page 4
1.a) one, two, b) one, one, two, c) one vertical line, 2 horizontal lines and 2 angles shaded
d) A with one vertical line and 2 angles shaded e) true
f) one vertical line, 2 horizontal lines and 2 angles shaded
2.a) 3 b) one sector dotted c) true, yes, horizontal
d) vertical lines e) A, start f) 3 equal sectors, one sector dotted

Page 5
1.a) all squares b) direction of arrow
c) turning 45° clockwise
d)

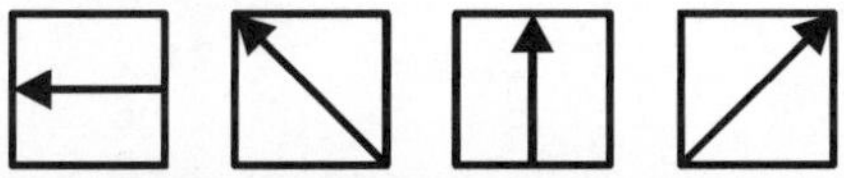

2.a) circle b) diameters c) 4, 5, 6
3.a) triangle b) size c) 1 cm, 1.5 cm, 2 cm, 1 cm
d) triangle 5 = 1.5 cm; triangle 6 = 2 cm.
4.a) number of oranges in each square: 1, 4, 9, 16, 25, 36, 49
b) square numbers, increasing difference of odd numbers (3, 5, 7, 9, 11, 13...) c) 140 oranges

Page 6
1.b) 3 c) 3 d) 9
2.a) red/green, brown/green, cream/green, fawn/green, red/orange, brown/orange, cream/orange, fawn/orange, red/red, brown/red, cream/red, fawn/red
b) 12 c) 15

Page 7
1.a) 5 white, black, navy 6 white, black, tan
7 white, white, navy 8 white, white, tan
9 blue, black, navy 10 blue, black, tan
11 blue, white, navy 12 blue, white, tan
b) 12 different outfits
2.b) red, black, navy; red, black, tan; red, white, navy; red, white, tan; blue, black, navy; blue, black, tan; blue, white, navy; blue, white, tan c) 8

Page 8
1.a) 300 kg b) 150 kg c) 42 + 39 + 69, 44 + 35 + 71 = 150 kg
d)

42	39	69
71	35	44

2.a) 9, 4, 36 b) check diagram c) 32
d) No the corner posts were counted twice

Page 9
1.a) 9 on one side 3 + 5 + 1 kg on other b) 5 kg = 3 kg + 2kg
c) 9 kg (LHS) 5 kg + 4 kg (RHS)
d) place 9 and 3 on LHS, 7 kg object + 5 kg on RHS
e) 9 + 5 = 11 kg + 3, 9 + 5 + 3 = 17 kg, 9 = 6 kg + 3
2.a)

11	14
9	← 16 →

b) Place 11 cm and 14 cm end-to-end.
Match the 9 with one end.
The difference is exactly 16 cm. c) picture

Page 10
1.a) L = 60 m, W = ½ length, 3 strands of wire
c) 30 cm d) 60 + 30 + 60 + 30 = 180 m
e) 3 f) 180 x 3 = 540 m
2.a) A → B = 29, A → C = 67, A → D = 132 km
c) 29, 38, 65 d) Byfield to Dawson e) add 38 and 65
f) 103 km g) 170 km. A → D = 132 D → E = 38
∴ A → E = 132 + 38 = 170 km

Page 11
1. 4 trips 2. 550 L 3. Carlos, Verne, June, Sharma
4. 30 m 5. 10 posts 6. Masa 141, Molly 140, Maryann 137, Mike 135 7. diamonds, platinum, gold, silver

Page 12

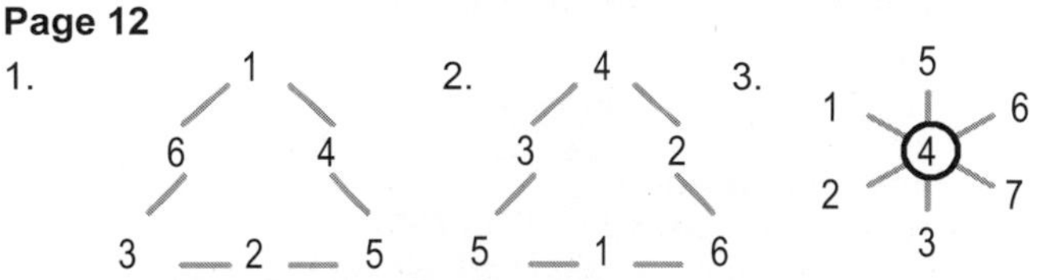

4. 6 in centre 1 + 11 one diameter, 2 + 10, 3 + 9, 4 + 8, 5 + 7 will give totals of 18

Page 13
1. 432 + 657 = 1089 is one possibility 2. 14
3.a) 3 b) 6 c) 10

Page 14
1.a) yes b) yes c) no d) no e) guess again
f) 16 knives, 14 forks, 17 spoons
2. 3 x 50 c, 2 x $1 and 1 x $2; or 5 x $1 and 1 x 50 c
3. 8 bicycles, 6 tricycles

Page 15
e) 28 and 19
2.a) 31, 32, 33, 34, 35, 36, 37, 38, 39
b) numbers following each other d) 33, 34, 35
3.a) 31, 32, 33, 34, 35, 36, 37, 38, 39
b) any ones, in any order c) 32, 37, 38 is one combination

Page 16
1.a) 24 b) 16 c) 8 d) correct
2.a) 48 b) double, 24 c) subtract d) 18
e) enter amount and work forwards
3.a) 47 b) subtract c) multiple d) dividing e) 6
f) work forwards
4. add 8 to 22 = 30, divide by 3 = 10, check by working forwards: 10 x 3 – 8 = 22

Page 17
1.a) □ x 6 + 32 = 98 b) subtract 32 from 98 = $66 c) $11
d) work forwards
2.a) 136 cm b) 141 cm c) subtract 3 cm d) metres
e) 1.38 m f) correct
3.a) Thursday $390 b) Wednesday $375 c) Tuesday $387

Page 18
1.a) 53 – 16 = 37 b) 37 + 29 = 66 c) 66 – 29 + 16 = 53
2.a) □ x 2 – 10 = 30 b) 20 c) work forwards
3.a) 8 b) 15 kg c) 12 m d) $20

Page 19
d) 3 days Broken Hill → Sydney - Broken Hill → Ivanhoe, Ivanhoe → Forbes 126 km less, Forbes → Sydney (53 less) 392 km
e) work backwards f) 392 km g) 53 less h) 445 km
i) 126 km more j) 445 – 126 = 319 km k) 319, 445, 392
l) addition m) 1156 km n) yes o) zero

Page 20
1.a) 1 b) zero c) 1
2.a) Sandy, Sam = 2 b) Sam 1 c) zero d) 3
3.a) 4 b) 3 c) 2 d) 1 e) 0 - already met others
f) 4, 3, 2, 1, 0 g) 10
4.a) 8 b) 7 c) 6 d) 5 e) 4 f) 3 g) 2 h) 1 i) zero
j) 8, 7, 6, 5, 4, 3, 2, 1, 0 k) 36

Answers cont.

Page 21

1.a) Dad, Jim, Jake, Sue, Andrea, Paula - 6
b) 6 c) 6 d) 6 e) 6 f) 3 x 6 = 18 g) 42
2.

3.a) paid $1.60 sold $2.00 profit 40c; bought $1.80, loss $1.40; sold $2.20 profit 80c b) profit $0.80

Page 22

1.a) 5 boys is minimum b) yes, yes, yes
2.c) try again d) 20c, 50c, 10c, $2, $1

Page 23

1.a) Arthur, Betty, Charles, Don, Ethel, Fred, George, Helen, Iris b) 1 2 3 4 5 6 7 8 9 c) Arthur, Charles, Ethel, George, Iris d) 10 11 12 13 14
e) Charles, George f) 14, 15 g) George
2.b)

A	B	C	D	E	F	G	H	I	B	D	F	H	B	F
1	2	3	4	5	6	7	8	9	10	11	12	13	13	14

c) Fred

Page 24

1.

3	2	2	2	1	1	1	1	1	0	0	0	0	0	0	0
0	2	1	0	4	3	2	1	0	6	5	4	3	2	1	0
0	0	2	4	0	2	4	6	8	0	2	4	6	8	10	12

2.a) 3 - one for each colour
c)

red	1 1 1 1 2 2 2 3 3 4
white	4 3 2 1 3 2 1 2 1 1
blue	1 2 3 4 1 2 3 1 2 1

d) 10 combinations e) 10 combinations f) same
3. yes, 2, 2, 2 only

Page 25

1.a)

1	2	3	4	5	6	7	8	9
2	4	6	8	10	12	14	16	18
9	8	7	6	5	4	3	2	1
36	32	28	24	20	16	12	8	4
10	10	10	10	10	10	10	10	10
38	36	34	32	30	28	26	24	22

c) 5 ducks 5 dogs
2.

50 -	2	1	1	1	0	0	0	0	0	0
20 -	0	2	1	0	5	4	3	2	1	0
10 -	0	1	3	5	0	2	4	6	8	10

a) 3 x 20c, 4 x 10c b) 5 x 20c c) 2 x 50c d) 10 x 10c e) no

Page 26

1. Big Burger $2

Foods		Drinks	
Big Burger	$2.95	Coffee	$1.50
Cheeseburger	$2.30	Soft Drink	$1.20
Chicken Burger	$2.80	Milkshake	$1.90
Chips	$1.50		
Wedges	$1.80		

b) $5.90, $4.60, $1.50, $3.60, $15.60, $4.40
c) Wedges, Chicken Burger, Milkshake
d) (i) no (ii) no change (iii) $1.50
e) $76.70, $23.00, $50.40, $31.50, $10.80, $7.50, $22.80, $24.70 total = $247.40

Page 27

b)

1	2	4	8	16	32	64	128	256	512	1024	2048
1	3	7	15	31	63	127	255	511	1023	2047	4195

c) $4194 d) $3600 e) doubling earns more
2.a)

time	1	2	3	4	5	6	7	8	9	10	11	12
sheets	9	18	27	36	45	54	63	72	81	90	99	108

b) (i) 54 pages (ii) 3 min (iii) 5 x 6 min (iv) 54 pages
(v) yes 99 (almost 100) takes 11 minutes; 500 = 5 x 99 so 5 x 11 min = 55 minutes, less than 1 hr

Page 28

1.a)

B	C	D	E	F	G	H	I	J	= 9
C	D	E	F	G	H	I	J	= 8	
D	E	F	G	H	I	J	= 7		
E	F	G	H	I	J	= 6			
F	G	H	I	J	= 5				
G	H	I	J	= 4					
H	I	J	= 3						
I	J	= 2							
J	= 1								
0									

b) 45
c) A shook with B. This is not a new one, but a reverse of the other.
2.a) make list
b)

manual V6 red	auto V6 red
manual V6 black	auto V6 black
manual V6 white	auto V6 white
manual V8 red	auto V8 red
manual V8 black	auto V8 black
manual V8 white	auto V8 white

c) 12 combinations e) yes

Page 29

1.a) list could be in order of months or in sales ascending/descending
b) January c) 31 more d) 3 less e) 375
f) January, May, June, April, February, July, March
2.a)

M	$198	F	$372	Th	$102
T	$265	T	$265	M	$198
W	$248	W	$248	W	$248
Th	$102	M	$196	T	$265
F	$372	Th	$102	F	$372

b) Friday c) Monday d) $1185 e) $107
f) Thursday - poorest sales

TEST 1 - 4 answers on Page A4

Page 34

1.a) plane b) 3 c) 4 d) 5 e) counting numbers
f) 6, 7, 8 and 9
g)

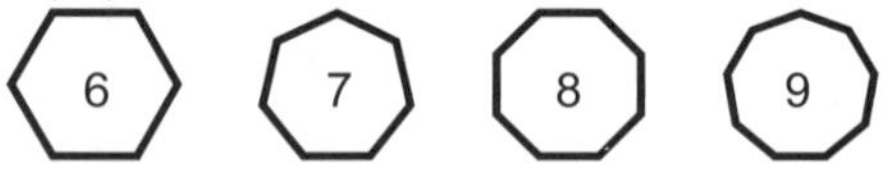

h) hexagon, heptagon, octagon, ennagon
2.a) squares b) 1 c) 4 d) 9 e) 1, 4, 9 f) square
g) 16, 25, 36
h)

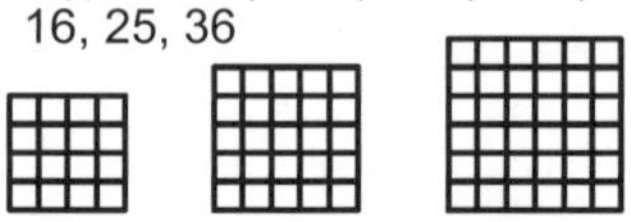

i) 64 j) 10th h) 12th
l) no - 200 is not a square number

Page 35

1.a) 1, 3, 6, 10 b) triangular numbers c) 15, 21, 28
2.a) 1, 3, 5, 7, 9 b) odd numbers
c) 11, 13, 15 d) number in top line, 1 in middle, same number in bottom line as in top
3.a) 3 b) 7 c) 15 d) 3 + 4 = 7, 7 + 8 = 15
e) 15 + 16 = 31, 31 + 16 = 63
f) (i) 2 (ii) 4 (iii) 8 g) 2, 4, 8
h) 16, 32 → 15 + 16 = 31, 31 + 32 = 63 i) SAME

Page 36

1.a) whole b) odd c) triangular d) even e) square f) prime
2.a) 70, 72, 74 b) 201, 199, 197 c) 144, 169, 196 d) 128, 64, 32
3.a) 111 b) 222 c) 333 d) 3 digits the same
e) 444 - correct f) 555, 666, 777, 888, 999 g) 777, 999
h) no, 1110 i) 1110, 1221, 1332, 1443, 1554, 1665, 1776, 1887, 1998

Page 37

1.a) 0.1111111 b) 0.2222222 c) 0.3333333
d) 0.4444444, 0.5555555, 0.6666666, 0.7777777
e) 0.8888888 f) false 9 ÷ 9 = 1

Page 37 cont.

g) 1.2222222, 1.6666666, 1.7777777, 1.4444444, 2.3333333, 2

2.a) 0.3434343, 0.3737373, 0.7575757, 0.3535353, 0.3838383, 0.6666666, 0.3636363, 0.3939393, 0.1212121

b) yes, the two-digit number is repeated as a decimal

c) 0.4040404, 0.6363636, 0.2222222

d) 0.0101010, 0.0202020, 0.0303030, 0.0404040, 0.0505050, 0.0606060, 0.0707070, 0.0808080, 0.0909090, 0.1010101

3.a) 0.0909090, 0.1818181, 0.2727272, 0.3636363

b) 0.4545454, 0.6363636, 0.8181818, 0.9090909

d) No. Any number divided by itself = 1

Page 38

b) 36 c) 50 g instant, 50 g decaf, 50 g freeze, 50 g espresso; 100 g instant, 100 g decaf, 100 g freeze, 100 g espresso; 200 g instant, 200 g decaf, 200 g freeze, 200 g espresso

d) 48

2.b) 16 cars

Page 39

1.a) check drawing b) 24 meals

2.a) check drawing b) true 18 combinations

3.a) check drawing b) 72 options

Page 40

c) ¼ d) 60 litres e) ¼ = 60 L then whole = 240 L

f) No. How much to be added to fill the tank? g) ¼

h) 60 litres

2.e) Mad Marty, Booker, Missy, Flying Star, Fine Finish

Page 41

1.a) check drawing b) $\frac{3}{8}$ c) 42 L d) $\frac{3}{8}$ = 42 L, $\frac{8}{8}$ = 112 litres

2.b) $10.50 c) ¼ d) $42 e) correct

3.c) 0.5 d) 25 km e) 50 km

f) enter data into problem, work forwards

Page 42

1.b) No. convert to same type c) 0.4 + 0.1 + 0.2 = 0.7

d) 0.3 e) 0.3 f) 10 tenths = $120 g) work forwards

2.a) match ends of 32 cm and 47 cm. Difference is 15 cm

b)
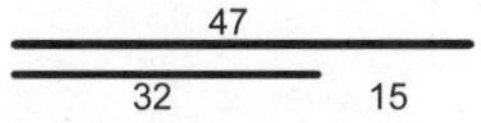

c) most prefer diagram

di)
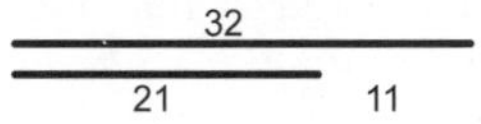

ii)
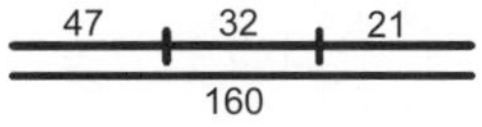

iii)

iv)
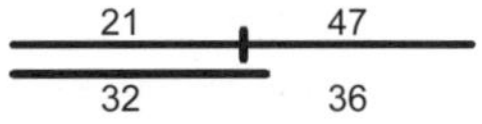

v)
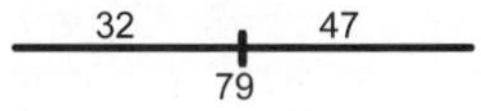

vi)
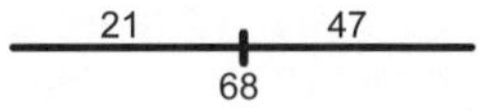

Page 43

1.b) 50 cm c) 35 cm d) 85 cm e) 70 cm f) 120

g) 105 h) 190 i) 175 j) 225 - out

2.b) his gain per minute is 10 m in 5 m out = 5 m gain/minute ∴ 100 m will take 20 minutes c) 20

Page 44

1. 6 x 5-pointed, 7 x 7-pointed 2. 3 x $3, 21 x $5

3. 16 x $3, 8 x $5

Page 45

1.a) 10 x 40 c, 16 x 50 c b) work forwards

2.b) 20 adults 42 children

c) No. number of children over adults d) 22 children

3.b) find number of cars c) 18 cars

Page 46

1. 32 and 13 2. 52 and 30 3. 16

Page 47

1. 35 2. 18 3. $18 4. 25 kg

Page 48

1.a) total mass 168 kg b) Amy 38 kg c) Amy 38 kg

d) 3 kg lighter e) 35kg f) 38 kg, 53 kg because he is 15 kg heavier g) 42 kg, 7 kg heavier than 35 kg

h) 38 kg, 42 kg, 35 kg, 53 kg i) no, should be lightest to heaviest j) Carla 35 kg, Amy 38 kg, Bruno 42 kg, Dominic 53 kg k) (i) 168 kg (ii) 38, 42, 35, 53 (iii) 168 kg (iv) yes

2.a) Stack C = $15 b) D = $10 c) $10, $20 d) $20, $40

e) A = $40, B = $20, C = $15, D = $10 f) no

g) $85, a total of amounts is required

Page 49

1.a) grey = 10 b) add 10 and 22 c) 32 blondes

d) false, true e) 16 brunettes f) 10 g) total in group

h) 68 i) correct

2.a) X = 180 L b) Y = 90 L c) Z = 300 L d) W = 220 L

e) total volume of all 4 f) 790 litres

g) work forwards and check

Page 50

1.a) 3rd, 4.75 m b) 468 cm c) 4.92 m d) 4.27 m

e) 4.27, 4.92, 4.75, 4.68 f) check circling g) yes

2.a) $250 million b) Janet's c) $222 million more

d) $472 million e) $5840 million f) adding $565 million

g) $6405 million h) Kerry's fortune in billions

i) ÷ by 1000 j) $6.405 billion

Page 51

1.a) Jim arrived 7.05; Jenny: 7.05 - 2 min = 7.03; Jane: 7.03 - 2 min = 7.01, Jack: 7.01 - 2 min = 6.59

b) varied answers c) 1st Jack 6.59, 2nd Jane 7.01, 3rd Jenny 7.03, 4th Jim 7.05

2.c) 692 km d) 361 km, 1052 km, 692 km e) 2105 km

f) yes g) answer is correct

Page 52

1.a) 10 b) 9 c) 8 d) 7 e) 6 f) Five g) 4 h) 3

i) eighth j) 9, 1 k) no one l) 9, 8, 7, 6, 5, 4, 3, 2, 1, 0

m) 45 n) descending counting numbers

2.a) more b) more people d) people 13, 12 and 11

e) 12, 11 and 10 farewells g) 33 + 45 = 78 h) 13 to 91

Page 53

1.a) 3 + 1 b) 4 + 1

c) 5 + 1 = 6, 6 + 1 = 7, 7 + 1 = 8, 8 + 1 = 9

d) Yes. one less than the person's number to remember then add own e) remember 14 and add 1

g) remember 24 and add 1

2.a) Amounts in cents: 1, 2, 4, 8, 16, 32, 64, 128, 256, 512, 1024, 2048, 4096, 8192, 16 483, 32 768, 65 536, 131 072, 262 144, 524 288, 1 048 576

b) doubling much better for Nguyen. $50 compared to $10 485.76

Page 54

1.a) 53, 56, 59, 62, 65, 68 b) 54, 58, 62, 66 c) 62

2.d) $5, $50, $100, $10, $20

3.a)

b) 91 c) 285 d) 11 layers

Page 55

1.c) M, M, W, M, W, W

2.a) $8 b) $3 more = $11 c) Betan x 2 = Bernice

d) $22 e) $27 f) Barry $8, Bernice $22, Bob $27, Betan $11

3.a) guessing and checking is one way c) 9, 15, 7, 8

Page 56

1.a) 10 combinations b) no c) 6 combinations d) no

e) some miss target completely

2.a) check grid b) 7

c) 6 combinations: 6 + 1, 5 + 2, 4 + 3, 3 + 4, 2 + 5, 1 + 6

d) 2 and 12

e) each die must score at least 1 - smallest number is 2

f) 4: 3 + 6, 5 + 4, 4 + 5, 6 + 3. 6 + 3 appears to be the same as 3 + 6 but the numbers appear on different dies

Answers cont.

Page 57

1.a) 1 b) 2 whiting, 7 bream c) 10 fish d) 2, 4, 8 = 14 fish
e) 18 bream, 24 whiting, 12 flathead = total 54
f) 6 flathead, 12 whiting, 30 fish
2.a) no b) no c) 60 x 1, 30 x 2, 20 x 3, 15 x 4, 12 x 5, 10 x 6, 6 x 10, 5 x 12, 4 x 15, 3 x 20, 2 x 30, 1 x 60 = 12 combinations
d) 6 x 12 = 72 too many chairs e) 9 combinations

Page 58

1.a) 1 59c, 2 $1.18, 3 $1.77, 4 $2.36, 5 $2.95, 6 $3.54, 7 $4.13, 8 $4.72
b) add cost of 8 + 4 ($4.72 + $2.36 = $7.08) or double 6 ($3.54 x 2) = $7.08 or...
c) 5 cost $2.95, 50 = 5 x 10, so cost is $2.95 x 10 = $29.50
d) final adding $4.13 + $4.72 = $8.85 not $8.75
e) 14 (7 = $4.13 so $8.26 must = 14)
2.a)

1	2	3	4	5	6	7	8	9	10	11
35	70	105	140	175	210	245	280	315	350	385
72	144	216	288	360	432	504	576	648	720	792

b) (i) 9 minutes (ii) 280 toy cars
c) 16 minutes (8 x 2) = 560 toys (280 x 2) = $11.52 ($5.76 x 2) d) $12.96 ($6.48 x 2) = 630 cars (315 x 2) = 18 minutes (9 x 2) e) 665 toys (315 + 350) = 19 (9 + 10) minutes = $13.68 ($6.48 + $7.20)

Page 59

1. 29 + 17 2. 18 and 23 3. 9 x 75 c and 8 x 45 c
4. 12 wrong answers 5. Esmeralda 106, Agnes 53 years, Sadie 73 years 6. 75 and 36

Page 60

1.a) 124 and 126 b) no c) 142, 146 d) 162, 164
e) 214, 216, 241, 246, 261 and 264 f) 412, 416, 421, 426, 461, 462 g) 612, 614, 621, 624, 641, 642.
2.a) AMC, AMP, AMG, ACP, ACG, APG, MCP, MCG, CPG, PGM
b) 10 combinations

Page 61

1.a) 3 x 50 + 20 + 10; 2 x 50 + 4 x 20; 3 x 50 + 3 x 10; 2 x 50 + 3 x 20 + 2 x 10; 2 x 50 + 2 x 20 + 4 x 10; ... b) 24 ways
2.b) plus 1, 6, 8; 1, 6, 9; 1, 6, 10; 1, 7, 8; 1, 7, 9; 1, 7, 10; 2, 4, 8; 2, 4, 9; 2, 4 10; 2, 5, 8; 2, 5, 9; 2, 5, 10; 2, 6, 8; 2, 6, 9; 2, 6, 10; 2, 7, 8; 2, 7, 9; 2, 7, 10; 3, 4, 8; 3, 4, 9; 3, 4, 10; 3, 5 8; 3, 5, 9; 3, 5, 10; 3, 6, 8; 3, 6, 9; 3, 6, 10; 3, 7, 8; 3, 7, 9; 3, 7, 10.
There are 36 different combinations.

Tests

Page 30 - Test 1

1.a) 53, 58, 63
b) ....., ..., ..., ...
c) square, pentagon, hexagon d) 1602, 1502, 1402
2.a)

1	2	3	4	5	6	7	8
7	14	21	28	35	42	49	56

b) 5 years c) 70
3.a) check diagram
b) white ham, white salad, white cheese, white vegemite, brown ham, brown salad, brown cheese, brown vegemite, wholemeal ham, wholemeal salad, wholemeal cheese, wholemeal vegemite
c) 16 d) pita ham, pita salad, pita cheese and pita vegemite

Page 31 - Test 2

1.a) 16, 15 spaces means 16 posts (1 to start, 1 to finish and 14 in between) b) 90 m, 30 m long 3 rails 3 x 30 = 90 m
2.a)

b) False: corner ones counted twice. 21 is correct total
3.a) 5 b) 22 cups, 28 saucers, 24 plates c) 63 bikes, 11 trikes

Page 32 - Test 3

1. red, blue, yellow, black, white 2. $60
3. 27 + 8 = 35 ÷ 5 = 7 (working backwards)
4.a) 28 greetings b) 12 people 5. 3
6.

7.a) 12 b) 6 c) 55

Page 33 - Test 4

1.a) 16
b) Sungsam 34 cm, 51 cm, 59 cm, 68 cm; Pannersonic 34 cm, 51 cm, 59 cm, 68 cm; Cony 34 cm, 51 cm, 59 cm, 68 cm; Haki 34 cm, 51 cm, 59 cm, 68 cm
2.

10	9	7	5	3	1	4	2	
20		1	2	3	4		1	2
50						1	1	1

3.

a)		b)		c)	
Monday	17	Wednesday	8	Sunday	35
Tuesday	19	Thursday	12	Friday	21
Wednesday	8	Saturday	14	Tuesday	19
Thursday	12	Monday	17	Monday	17
Friday	21	Tuesday	19	Saturday	14
Saturday	14	Friday	21	Thursday	12
Sunday	35	Sunday	35	Wednesday	8

d) chronological order e) worst to best sales f) best to worst sales g) 126 h) $157.50 i) 80 c j) $100.80

Page 62 - Test 1

1.a) 15 black triangles b) 21 black triangles c) 36 small triangles in large triangular shape d) yes - 10 small triangles along the base e) not in large triangular shape
2.a) 0.4444444 b) 0.3535353 c) 0.9090909
3.a) check diagram b) neck gold, neck silver, neck platinum; bracelet gold, bracelet silver, bracelet platinum; band gold, band silver, band platinum c) 9 combinations
4.a) 15 rings b) check diagram

Page 63 - Test 2

1. 32 years, 11 years 2. 9 buses minimum 3. 62 metres
4. 85 marbles 5. 4 x 3 legs, 12 x 4 legs 6. 1050 kilolitres

Page 64 - Test 3

1. L = 147 cm W = 0.49 m H = 0.98 m 2. 72 cards (36 exchanges of 2 cards) 3. 8 steps 4. Tuesday
5. Green, Yellow, Orange, Red 6. 24 minutes

Page 65 - Test 4

1. 9 o'clock
2.a) AKQ, AKJ, AKIO, AQJ, AQIO, AJIO, KQJ, KQIO, QJIO
b) 9
3.a)

1	2	3	4	5	6	7	8	9	10	11
2	4	6	8	10	12	14	16	18	20	22
11	10	9	8	7	6	5	4	3	2	1
33	30	27	24	21	18	15	12	9	6	3
12	12	12	12	12	12	12	12	12	12	12
35	34	35	32	31	30	29	28	27	26	25

b) 10 bikes 2 trikes c) 30 d) you must have at least 1 bike as the maximum number of wheels cannot exceed 35 e) 12 tricycle wheels

Year 5:

Test 4

Lists and Tables will help you here.

1. Hardly Normal stocks 4 brands of TVs - Sungsam, Pannersonic, Cony and Haki in 4 different screen sizes: 34 cm, 51 cm, 59 cm and 68 cm.
The salespeople want to put up a display showing every TV they sell.

a) There must be enough space for ____ TV sets.

b) Make a list of all of the TVs that will go on display. ______________________________

____

____

2. Malcom has to give 90 cents change. He has 10 c, 20 c, and 50 c coins in the till.
On this table show all the possible ways he could give this amount of change.

3. Jenny's Cake Shop sells jam and cream doughnuts. Here are the sales figures for last week:
Monday 17, Tuesday 19, Wednesday 8, Thursday 12, Friday 21, Saturday 14, Sunday 35.
Create 3 separate listings for this data and explain the reason for the order you chose.

a) List 1	b) List 2	c) List 3

d) List 1 shows ____.

e) List 2 is arranged to show ____.

f) The third list shows ____.

g) What was the total sale of jam and cream doughnuts for this week? ____

h) Jenny sells her doughnuts for $1.25 each. Use your calculator to work out her sales figures for this week. ______

i) Each doughnut costs 45 c to produce. Her profit on one is______.

j) This means that her profit on ____ jam and cream doughnuts for this week was $_____.

Finding a Pattern

Find the pattern within these shapes and solve the problem.

1. 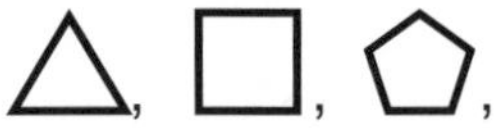 ______, ______, ______, ______

a) These shapes are plain / plane figures.

b) How many sides has the first shape? ____

c) There are ____ sides in the second.

d) The third shape has ____ sides.

e) Describe the pattern. ____

f) The next four shapes should then have ____, ____, ____ and ____ sides respectively.

g) Draw these shapes in the spaces provided.

h) The Greek root 'enna' means 9, 'octa' means 8, 'hepta' means 7, 'hexa' means 6, 'penta' means 5 and the root for side is 'gonos'. So a five-sided figure (the third in the series) is a pentagon.

Write the names for the other shapes you drew. ________, ________, ________, ________.

2.  __________, __________, __________

a) The above shapes are all __________.

b) The first square is made up of ____ small squares.

c) The second shape has ____ small squares.

d) How many small squares are in the third element in the series?

e) Write these three numbers ____, ____, ____.

f) The most interesting fact is that these numbers are __________ numbers.

g) Write the next 3 numbers in this series _____, _____, _____.

h) Complete the drawings in the series.

i) The 8th element would be a large square containing ____ small squares.

j) Which element would contain 100 small squares? ____

k) A large square divided into 144 small squares is the _____ element.

l) Would a square containing 200 small squares be included in this series? YES / NO.

Why? ____

____

Activities

Use similar processes to solve these shape patterns.

1. 

a) Write the first 4 numbers ____, ____, ____, ____.

b) Describe the series. ____

c) Complete the number series and draw the shapes. ____, ____, ____

2. 

a) The number of squares used is ____, ____, ____, ____, ____.

b) What sort of numbers are these? __________.

c) Predict the numbers for the next 3 elements ____, ____, ____.

d) Describe the pattern in which they are arranged and complete the drawings. __________

____

3. Check out these flights of birds.

a) There are ____ in the first flight.

b) ____ birds are in the second.

c) How many birds in the third flight? ____

d) Let's look at these numbers 3, 7, 15. 3 + ☐ = 7, 7 + ☐ = 15.

e) Predict the next two elements. ___, ___.

f) Another way to look at this is:

(i) There are ____ birds on each side of the flight in element 1.

(ii) How many on each side of the flight in the second? ____

(iii) Count the number on each side in flight 3. ____

g) Write these numbers ____, ____, ____.

h) Predict the next two elements. ___, ____.

i) Compare your two sets of predictions: SAME / DIFFERENT.

> There's often more than one way to solve a problem.

Finding a Pattern

Check out these number patterns.

1. Match up these series with their correct names.

a) 0, 1, 2, 3, 4, 5, 6, ______________

b) 1, 3, 5, 7, 9, 11, ______________

c) 1, 3, 6, 10, 15, 21, ______________

d) 2, 4, 6, 8, 10, 12, ______________

e) 1, 4, 9, 16, 25, 36, ______________

f) 2, 3, 5, 7, 11, 13, 17, ______________

prime	odd
even	square
triangular	whole

2. Add three more elements to each of these series.

a) 62, 64, 66, 68, ____, ____, ____

b) 209, 207, 205, 203, ____, ____, ____

c) 64, 81, 100, 121, ____, ____, ____

d) ____, ____, ____, 16, 8, 4, 2, 1

Let's look at these multiplying patterns.

3.

37 x 3 =
37 x 6 =
37 x 9 =
37 x 12 =
37 x 15 =
37 x 18 =
37 x 21 =
37 x 24 =
37 x 27 =

a) Use your calculator to do 37 x 3. Record your answer. Interesting, eh?

b) Calculate 37 x 6. Add this to the list. An idea forming perhaps?

c) Let's check 37 x 9. Can you see the pattern to the series?

d) Describe this pattern.______________________________

e) Predict 37 x 12. ____ Check it out.
Prediction was CORRECT / INCORRECT.

f) Predict the other elements ____, ____, ____, ____.

g) We had better check just to make sure (use your calculator).

37 x 21 = ____ and 37 x 27 = ____.

h) If this is true, then 37 x 30 must equal 1111. Do you agree? YES / NO.
Check it out on your calculator. Was it correct? YES / NO.

(i) Could there be a different pattern emerging? Check these out with your calculator. Predict the answers when you find the pattern.

37 x 30 = ____	37 x 39 = ____	37 x 48 = ____
37 x 33 = ____	37 x 42 = ____	37 x 51 = ____
37 x 36 = ____	37 x 45 = ____	37 x 54 = ____

Activities

Dividing patterns can be interesting too.

1.

a) Enter 1 on your calculator ÷ then 9 then = . Record the answer. __________

b) 2 ÷ 9 = ☐ shows on the screen as __________

c) 3 ÷ 9 = ☐ shows on the screen as __________

d) See the pattern. You predict these 4 ÷ 9 = __________ 5 ÷ 9 = __________

6 ÷ 9 = __________ 7 ÷ 9 = __________

e) Use your calculator again for 8 ÷ 9, just to make sure you're on track. 8 ÷ 9 = __________

f) If this is true, then 9 ÷ 9 = 0.99999999. Is this TRUE / FALSE? Why? __________

____

g) Let's go on. Predict and check (calculator) these:

11 ÷ 9 = __________ 15 ÷ 9 = __________ 16 ÷ 9 = __________

13 ÷ 9 = __________ 21 ÷ 9 = __________ 18 ÷ 9 = __________

2.

a) 34 ÷ 99 = __________ 37 ÷ 99 = __________ 75 ÷ 99 = __________

35 ÷ 99 = __________ 38 ÷ 99 = __________ 66 ÷ 99 = __________

36 ÷ 99 = __________ 39 ÷ 99 = __________ 12 ÷ 99 = __________

b) Is there a pattern? YES / NO. Describe this pattern. __________

____

c) Predict these: 40 ÷ 99 = __________ 63 ÷ 99 = __________ 22 ÷ 99 = __________

d) Predict the following answers when these numbers are divided by 99.

1 __________, 2 __________, 3 __________, 4 __________, 5 __________,

6 __________, 7 __________, 8 __________, 9 __________, 10 __________

e) Check these with your calculator to prove you are right.

3. Investigate the pattern to be found when numbers are divided by 11.

a) Calculate: 1 ÷ 11 = __________ 2 ÷ 11 = __________ 3 ÷ 11 __________

4 ÷ 11 __________

b) Predict: 5 ÷ 11 = __________, 7 ÷ 11 = __________, 9 ÷ 11 = __________,

10 ÷ 11 = __________

c) Check your predictions.

d) Will 11 ÷ 11 equal 1.00100100? YES / NO. Why? __________

____

Compare the pattern ÷ 9 with the ÷ 11 pattern. Are there similarities / differences?

Tree Diagrams

This is a useful strategy to find combinations.

1. There are three brands of coffee on the shelves in the supermarket. All three brands come in 3 sizes (50 g, 100 g and 200 g). In each size there are four types (instant, decaf, freeze-dried and espresso). If one of each brand, size and type were selected, how many jars would be chosen?

a) Complete this Tree Diagram.

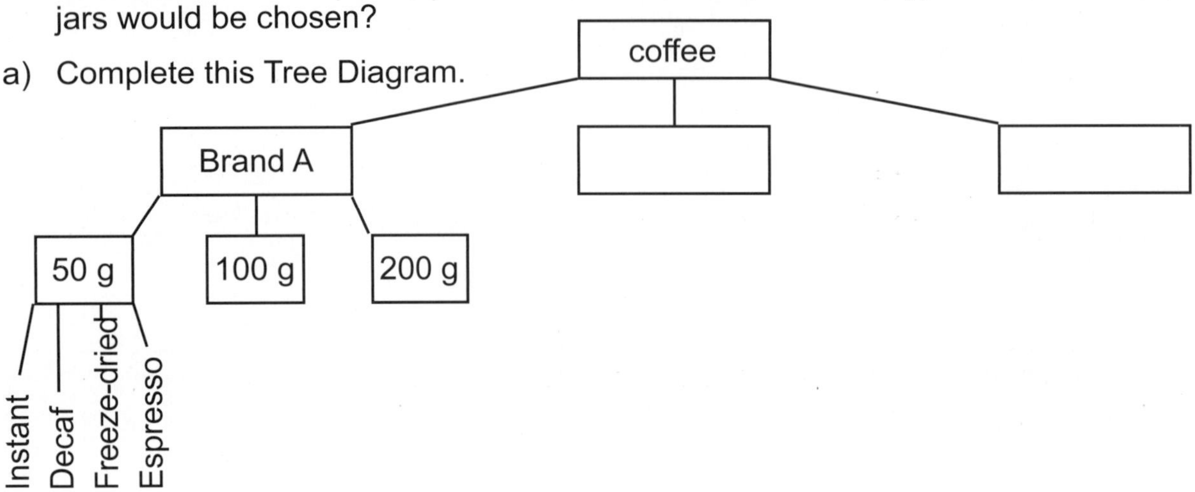

b) Count up the number of possible combinations. ____

c) Make a list of all the jars used by Brand B. ________________________________

____

d) If the supermarket owner decided to add another size for each brand and type, the number of jars would increase to ____.

Tree diagrams can also be drawn horizontally. Check this out.

2. Burn Motors sells new cars: Holdens and Toyotas in sedans, utes, station wagons and 4WDs. All have the option of manual or automatic. To display one of each in the sales yard, there must be space for ____ vehicles.

a) Complete the tree diagram.

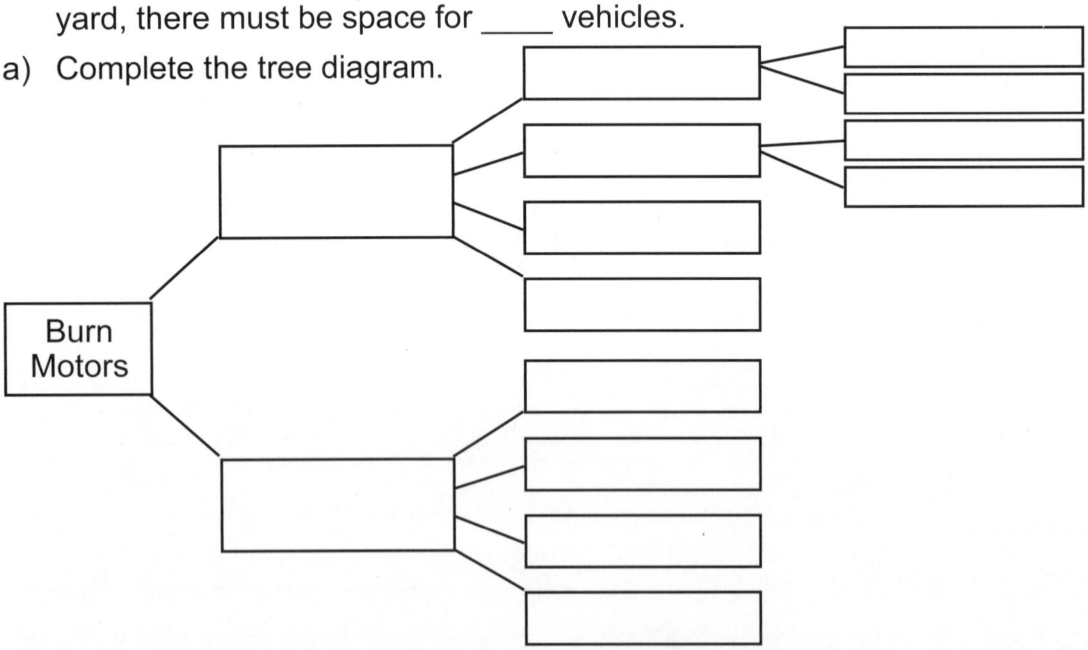

b) There must be enough space for ____ vehicles.

Activities

1. The Dine Inn restaurant is offering the following deal for a three-course meal – the choice of 3 entrees, 4 main meals and 2 desserts all for $11.95. How many different meals (by changing at least 1 course) can be created with these options?

a) Draw your tree diagram here.

b) ____ different meals can be created.

2. Mrs Ryan is travelling overseas. She's purchased clothing that she can mix and match – 3 blouses, 3 skirts and 2 jackets. Her trip will last 12 days. Has she enough to create a different outfit by changing at least one item for each day on her trip?

a) Show the combinations (tree diagram) here.

b) She has enough for the 12 days. TRUE / FALSE

3. Calculate the number of options for this computer – 333 MHz, 450 MHz or 500 MHz processor; 64 Mb, 128 Mb or 256 Mb of RAM; 4Mb or 8Mb Graphics card; 8Mb or 16Mb Sound Card, 15" or 17" monitor.

a) Show the options on a tree diagram.

b) In all there are ____ options.

Drawing a Diagram

Drawing a sketch or diagram can help you SEE a problem more clearly.

1. A water tank is half full. 60 litres were added. The tank is now three-quarters full. How much will have to be added to completely fill the tank?

Let's make a rough drawing of a tank and enter the data we know from the problem.

a) Mark in the half-way level.

b) Show the extra 60 litres added and the fraction now.

c) The difference (fraction) between half way and three-quarters is ____.

d) This fraction is equal to ____ litres.

e) If ____ is equal to 60 litres, then the whole tank must hold ____ litres.

f) Is this the answer to the question? YES / NO. Why? ______________________________

____

g) What fraction is still to be added? ____

h) The volume to be added is ____ litres.

2. Here are the results of last Saturday's fifth race. Mad Marty was 7 lengths ahead of Missy, Missy one length ahead of Flying Star, Brooker 5 lengths ahead of Fine Finish, Fine Finish 3 lengths behind Missy. Flying Star was not last. Work out the finishing order of these horses.

Let's use a line for the race and make quick sketches of the horses one at a time.

a) Show the position of Missy, including the distance.

b) Now draw Flying Star.

To enter the next horse, read the text until you find a reference to one of the horses already positioned.

c) Show Fine Finish's position.

d) Now add the other horses in the race.

e) Write the names in order – first to last. ______________________________

____

Activities

1. The fuel tank in Dad's 4WD was a quarter full. At the garage Dad added 42 litres. It is now $\frac{5}{8}$ full. The tank in this 4WD has a capacity of ____ litres.

a) Draw the diagram and label it.

b) What is the fractional difference between $\frac{1}{4}$ and $\frac{5}{8}$?

c) This fraction is equal to ____ litres.

d) If ____ (fraction) equals 42 litres,

then $\frac{8}{8}$ (full tank) is equal to _______.

2. Tony spent a quarter of his money on a computer magazine. He then spent half of what was left on a new CD. When he checked his wallet, he still had $10.50. Calculate how much he had at the beginning.

a) Draw the diagram and enter the facts.

b) How much did he have left after

spending on the CD?____

c) What fraction does this represent? ____

d) Use that fraction to find the whole

amount. ____

e) Check your answer by working the

problem forwards. Your answer is CORRECT / INCORRECT.

3. One vehicle set out from Aytown and travelled towards Beeville. A second vehicle left Beeville at the same time, heading for Aytown. After the first vehicle had covered 0.3 of the distance, the second had covered two-tenths of the distance and there was still 25 km separating the two vehicles. How far is it from Aytown to Beeville?

a) Draw the road from Aytown to Beeville.

b) Mark on the facts given in the problem.

c) Calculate the fraction separating the two vehicles. ____

d) This fraction is equal to ____ kilometres.

e) Now calculate the whole distance from Aytown to Beeville.

f) How will you check your working?____

____

Drawing a Diagram

1. Gerry spends 40% of his weekly wage on car repayments, one tenth of his money on board, 0.2 on clothing and spends the remaining $36 on entertainment.
What's Gerry's total income?

a) Label this diagram showing how much Gerry spends.

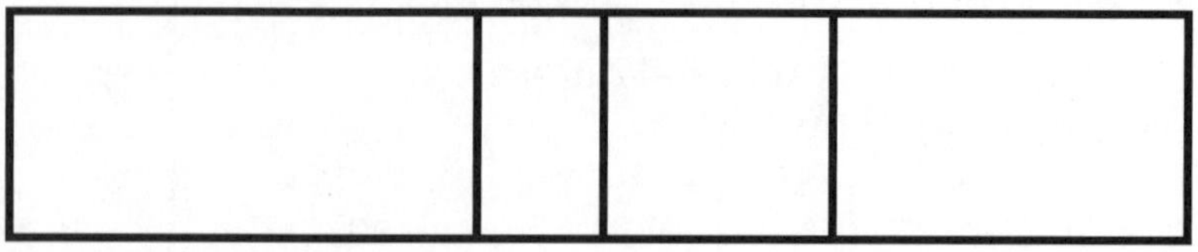

b) Can 40%, $\frac{1}{10}$ and 0.2 be added as they are? YES / NO.

What has to be done? ______________________________

c) Make the conversions and find the total. ______________________________

d) This means that the entertainment is equal to what fraction? ____

e) That fraction, ____, is equal to $36.

f) Calculate the whole amount. $____

g) Check your working. How will you do this? ______________________________

____

2. Sammy has 3 pieces of string: A 21 cm, B 32 cm, and C 47cm.

a) Describe how Sammy could show an exact length of 15 cm. ______________________________

____

b) Instead of words, show the same information in a diagram.

c) In your opinion, which is the easier to see/understand? __________

d) Draw diagrams showing the use of 2 or 3 pieces of string to show these lengths.

(i) 11 cm

(ii) 100 cm

(iii) 58 cm

(iv) 36 cm

(v) 79 cm

(vi) 68 cm

Activities

1. Freddy the Frog was climbing up a mossy bank. Every minute he rested after hopping up 50 cm. During his one minute rest period he slid back 15 cm. How long will it take Freddy to reach the top of this 2 metre mossy bank?

You could show this as a series of diagrams - one for each minute or one diagram showing all the data as you work it out.

Here's space for your diagram(s).

a)

Track Freddy Frog's progress here.

The distance up the mossy bank after

b) 1 minute ____

c) 2 minutes ____

d) 3 minutes ____

e) 4 minutes ____

f) 5 minutes ____

g) 6 minutes ____

h) 9 minutes ____

i) 10 minutes ____

j) 11 minutes ____

If you can break a problem down into a series of small steps, then each step is quite easy, and the big problem isn't as big as it first appeared.

2. Guy was fishing. He had 100 metres of line out when he hooked a big fish. Every minute he wound in 10 metres of line, only to find that the fish had pulled back 5 metres during that time. At this rate, Guy will take ____ minutes to land his fish.

a) Show the data on a diagram.

b) Describe how you solved this problem. ______________________________

____

c) It took Guy ____ minutes to land the fish.

Guessing and Checking

Guessing and checking is a valid strategy to use to solve problems provided that you use the first guess and check to make your second guess. Wild guesses rarely give the correct answer.

1. In the far-off land of Dlan, their flag has 5- and 7-pointed stars. The total number of stars is 13. There are 79 points on these stars. How many of each kind of star are on this flag?

		1st guess	2nd guess	3rd guess	4th guess
5-pointed	number				
	points				
7-pointed	number				
	points				
Total stars					
Total points					

The answer is ____ 5-pointed stars and ____ 7-pointed stars.

2. In Dlan, they use \$3 coins and \$5 notes. I have 24 notes and coins; total value is \$114. Therefore, I have ____ \$3 coins and ____ \$5 notes.

		1st guess	2nd guess	3rd guess	4th guess
\$3 coins	number				
	value				
\$5 notes	number				
	value				
Total coins and notes					
Total value					

3. My sister, Camille, has 24 notes and coins also but her total value is \$26 less than mine. How many of each has she?

Create your own grid here.

Activities

1. Milly bought 26 novelties for her party. Some cost 40 c, others cost $0.50. In all, she spent $12. How many of each did she purchase?

	1st guess	2nd guess	
No. of 40 c novelties			
Value			
No. of 50 c novelties			
Value			
Total value			

a) She purchased ____.

b) Describe how you checked this. ____

____

2. There are 24 seats on the bus. A seat can accommodate either 2 adults or 3 children. If all the seats were filled with 62 passengers, how many more children than adults were on the bus? Draw your grid below.

a) There were _____ adults and _____ children.

b) Is this the answer? YES / NO. Why?____

____

c) The answer is _____.

3. There were trucks (6-wheel) and cars (4-wheel) in a line for the tollway booth. Jimmy counted 30 vehicles. Minnie counted 144 wheels. There were _____ cars in this line. Draw your grid below to record your guesses and checks.

b) What does the question ask? ____

c) The answer is _____.

Always check back to make sure that your answer is indeed the answer to what the question in the problem has asked.

Guessing and Checking

1. I added two numbers, the total was 45. When I multiplied these numbers, the answer was 416. The two numbers were ____ and ____.

Let's make a guess.	20 and 25	Add them = 45 ✓ Multiple them = 500 ✗
The multiplication is too high, so let's try again.	18 and 27	Add them = 45 ✓ Multiple them = 486 ✗

Getting close to 416, so continue reducing one number and increasing the other. Show your guesses here.

The answer is ____ and ____.

2. Number A – Number B = 22. Number A × Number B = 1560.

Number A = ____ Number B = ____

This time record your guesses and checks on a grid.

	1st guess	2nd guess	
Number A			
Number B			
A – B			
A × B			

3. I squared a two-digit number and subtracted 40. The answer was 216. What was that 2-digit number?

	1st guess	2nd guess	
Number			
Squared			
Subtract 40			
Answer			

This problem could be solved by working backwards. Use this strategy to prove your guess and check.

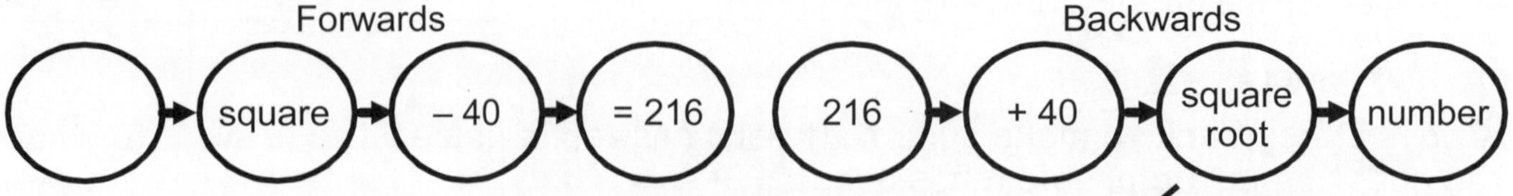

Use calculator here.

Activities

Use Guessing and Checking to solve these.
Prove your answers by Working Backwards.

1. A number to which 7 was added was divided by 3. If the result was 14, what was the number?

	1st guess	2nd guess			
Number					
Add 7					
Divide by 3					
Answer					

( ) → (+ 7) → (÷ 3) → (= 14) → ( ) → ( ) → ( )

2. □ – 18 + 23 = 23

	1st guess			
Number				
Subtract 18				
Add 23				
Answer				

( ) → (– 18) → (+ 23) → (= 23) → ( ) → ( ) → ( )

3. I had an amount of money. Mum doubled it. I spent all but one quarter, to leave me with $9. This means that I started with ____.

Amount			
Double			
One quarter			
Answer			

( ) → ( ) → ( ) → (= $9) → ( ) → ( ) → ( )

4. (□ kg + 45 kg) ÷ 2 = 35 kg.

Create your own table to record your guesses and checks.

Working Backwards

Using this strategy means finding the one fact that IS KNOWN and then working step-by-step through each UNKNOWN until the answer is found.

1. Altogether, 4 children have a total mass of 168 kg. Bruno is 7 kg heavier than Carla. Dominic is 15 kg heavier than Amy. The difference between Amy and Carla is that Amy (at 38 kg) is 3 kg heavier. Arrange these people in order (with their exact masses) lightest to heaviest.

There are two knowns. a)____________________________

List them. b)____________________________

c) Which of these knowns will lead you to the other masses? ________________

d) Carla is ____ kg _________ than Amy.

e) Therefore, Carla must have a mass of ____ kg.

f) If Amy is ____ kg then Dominic is ____ because ______________________

____

g) Having found Carla's mass, Bruno weighs ____ kg because __________________.

h) List the masses of the people: Amy ____kg, Bruno ____kg, Carla ____kg, Dominic ____kg.

i) Is this the answer required? YES / NO. Why?________________________

j) Write the correct answer.________________________________

k) Let's check our problem by using the other known.

(i) The total mass of all four children is ____ kg.

(ii) The individual masses are ____, ____, ____, ____.

(iii) The total of these four masses is ____ kg.

(iv) Do the two totals agree? YES / NO

2. A teller has four amounts of money (A, B, C and D). B is half that of A. D is half the amount of B. C is $5 more than D. There is a $10 and a $5 note in stack C. What is the total of the four amounts?

a) The known is __________________. b) This means that D is $____.

c) If D = $____ and D is half the amount of B, then B = $____.

d) A ÷ 2 = B and B = ____ so A must equal $____.

e) We have found A, B, C and D. List them. ______________________

f) Is this the required answer? YES / NO.

g) The answer is ____ because the question asks ______________________

____

Activities

1. In a group there are blondes, brunettes, dark-haired and grey-haired people. There are twice as many blondes as there are brunettes. There are 6 less dark-haired people than brunettes. If there were 22 more than the current ten grey-haired people, then the number of grey-haired and blonde would be the same. There are a total of ____ people in this group.

a) Find the only given known _____-haired = ____ people.

b) From this fact, tell how you will find the number of blondes. ______________________

____

c) The number of blondes is ____.

d) Which is true? Blondes × 2 = Brunettes TRUE / FALSE

Brunettes × 2 = Blondes TRUE / FALSE

e) Use this fact to find the number of brunettes. ____

f) Calculate the number with dark hair. ____

g) What does the question ask? ______________________

h) Answer that question. ______________

i) Check back through the problem putting in the numbers and making sure that your calculations are correct. Answer CORRECT / INCORRECT.

2. Describe each of the steps you take to solve this problem.

Sam has 4 partly-used drums of diesel fuel (Brands W, X, Y and Z). Z contains 80 L more than W. Y is 210 litres less than Z. X is double that of Y. The contents of X will completely fill an empty 180 L fuel tank. What's the total volume of all four drums?

a) ____

b) ____

c) ____

d) ____

e) What does the question ask? ______________________

f) The answer is ______________.

Describe how you checked your working and answer.

g) ____

____

____

____

Working Backwards

1. At the Olympics, these 4 jumps were recorded for the winner of the Long Jump. 1st attempt was 65 cm short of his second jump. The second jump was the longest. On the third try he jumped 4.75 metres. The fourth attempt was 7 cm shorter than the third. His last jump was only 24 cm short of his best on that day. Find the four distances jumped and circle the winning one.

a) Enter the known jump length here. __________

b) If the 3rd jump was _____, then the fourth was _____ cm.

c) Find the length jumped on the second attempt. _____m

d) Now work out his first jump. _____m

e) Record the results.

1st jump	2nd jump	3rd jump	4th jump

f) Circle the winning jump.

g) I have answered the question. YES / NO.

If a problem contains large numbers, don't allow this to confuse you. Use the same strategy to solve it.

2. John has a personal fortune which is $222 million more than Janet. Rupert has $5590 million more than Janet. Kerry is even richer, with $565 million more than Rupert. If Janet has a quarter of a billion dollars then Kerry has a total fortune of _____ billion dollars. (Note: $1 billion = $1000 million)

a) One quarter of a billion = _____ million.

b) Whose fortune is equal to this? ________________

c) John has $_________ more / less than this.

d) His fortune is equal to ________________.

e) Now calculate Rupert's wealth. ________________

f) From this fact you can calculate Kerry's total by ______________________________________.

g) Kerry has control of $_____ million.

h) What is the question asking?____

i) How will you convert millions to billions?____

j) Give your answer here. ____

____

Activities

1. Four friends (Jane, Jenny, Jim and Jack) were to meet outside the pool for swimming training. Jack was a minute early. Jane arrived 2 minutes after Jack and 2 minutes before Jenny. When Jim arrived 2 minutes later than Jenny, she told him he should have been there 5 minutes ago. If Jim was there at 7.05, list the people in order and the times of their arrival.

Show your working here.

a) ____

____

____

____

Show your answer two ways.

b) In words

____

____

____

____

c) As a table

	Person	Time
1st		
2nd		
3rd		
4th		

2. It is 361 km from Alston to Bedford. From Alston through Bedford through Carlton to Drayton is 2105 km. From Bedford, Carlton is 691 km further than the distance from Alston to Bedford. How far is Drayton from Carlton?

Drawing a diagram will help you SEE the problem.

a) Complete this.

b) Describe how you went about solving this. ________________________

____

____

c) Your answer is ________.

d) What were the distances? Alston to Bedford ____km, Bedford to Carlton ____, Carlton to Drayton ____.

e) What is the total of these 3 distances?________

f) Does this total agree with the given total? YES / NO

g) What does this prove?________________________________

Acting It Out

1. Ten friends were at a party. Towards the end, the first to leave said 'goodbye' to each of his friends and left. The second did the same. So did the third, fourth and so on until there was only one person (the person who owed the home) left. Just how many times was the word 'goodbye' used as these people departed?

Use counters to represent the people.

a) There were ____ people at the party.

b) The first to leave would have said 'goodbye' to the other ____ people.

c) The second said 'goodbye' to the other ____ people remaining.

d) How many did the third say 'goodbye' to? ____

e) The fourth said 'goodbye' to the ____ remaining.

f) _____ 'goodbyes' were said by the 5th person.

g) Number 6 to leave said it ____ times.

h) The seventh to depart said 'goodbye' to ____ people.

i) The ____ person to leave said 'goodbye' to 2 people.

j) The last to leave (number ____) had ____ people left to say 'goodbye' to.

k) Who did the tenth person have to say 'goodbye' to? _______

l) Go back and gather the numbers ____, ____, ____, ____, ____, ____, ____, ____, ____, ____.

m) Find the total. ____

n) Was there a pattern to these numbers? YES / NO. Describe it. ________________

____

2. At their next gathering the number swelled to thirteen. If each person said goodbye to all of those left as they departed, the number of farewells would increase to ____.

a) Will the number be more/less than the previous answer? ________

b) Why? ____

c) Predict (estimate) what the answer will be. ____

d) Which ones only have to be calculated? ________________________

e) Use the pattern to find these numbers. ________________________

f) Check your patterned answers by Acting It Out.

g) Add this total ____ to the previous question total ____ and this will give you a total of ____ farewells for 13 people.

h) They expect 14 at their next function. At that time, the number of farewells will increase by another ____ to make a total of ____.

Activities

1. Yesterday, in Milly's class, they tested their memories by playing the game 'I went to town and bought... ' Here are the first 3 players' answers.

Jane: 'I went to town and bought a new CD.'
Mike: 'I went to town and bought a new CD and a Mercedes Benz.'
Trish: 'I went to town and bought a new CD, a Mercedes Benz and a diamond ring.'

a) The fourth person had to remember ____ articles and add ____.

b) Number 5 remembered ____ and added ____.

c) Add the missing elements into this table.

d) Is there a pattern? YES / NO.

Describe it.______________________

____

e) Use the pattern to predict what the 15th person in line must do. ______________________

____

f) Check your prediction by Acting it Out.

Person	needs to remember	then adds	making a total of
6			
7			
8			
9			
10			

g) There are 24 children besides Milly in this class.

Milly was last in line, she had to ______________________.

2. Nguyen is a real brain when it comes to mathematics. His dad promised him $50 at the end of three weeks if he helped every day in their fruit shop.
Nguyen said, 'Rather than that, what about just one cent on the first day, two cents for the second and continue doubling just like that for the 3 weeks?'
His dad knew Nguyen was up to something so he decided to check it out.
Which would be the better deal for Nguyen?

Use counters to present the cents. Record your results on a table to keep things in order.

a)

Day	Amount	Day	Amount	Day	Amount
1	1 cent	8		15	
2	2 cent	9		16	
3	4 cent	10			
4					
5					
6					
7					

b) Doubling 1 cent or $50. Which is the better deal?______________________

Year 6: Unit 11

Acting It Out

1. Jenny has a big bunch of grapes. If she eats them 3 at a time, she will have two left over. If she eats 4 at a time, her last mouthful will again be 2 grapes. Looking at the bunch, she estimates that there are between 50 and 70 grapes. What is the exact number?

Use 70 or so counters/buttons to model this problem.

a) Find all of the numbers which will divide by 3 with a remainder of 2.

____

b) By modelling, find all of the numbers >50 <70 which, when divided by 4, have a remainder of 2. Write them here. ____________________________

c) Compare the two lists. Are any number(s) common to both lists? YES / NO.

Name it/them. ___________

2. There are 5 notes ($5, $10, $20, $50, $100) in a stack. The $50 is just under the $5 and just on top of the $100. The $5 is above both the $10 and $20. The $10 note is not on the bottom of the pile. What is the order of these notes from top to bottom?

Make up the five notes using pieces of paper.

a) Make an arrangement. Record it here. ____

b) Check it against the problem. Is your arrangement CORRECT / INCORRECT?

c) Record your second try. How did it check out? ___________________________________

d) The order is ______________________________.

Using these pieces of paper as models, you are able to use the Guessing and Checking strategy as you Act It Out.

3. Lex made a display of oranges in the form of a square-based pyramid at the Royal Easter Show. Here are some of the layers.

top layer	next layer	3rd layer	4th layer	5th layer	6th layer
●	●● ●●	●●● ●●● ●●●			

a) Use tennis balls to model the next layer. Record it on the table above.

b) How many tennis balls will you need to create a model of this pyramid with 6 layers? ______

c) For 9 layers you'll need _____ balls.

d) Given 500 balls, you could make a pyramid with _____ layers.

Activities

1. In line at the checkout there are six people (3 men and 3 women). Two men are chatting as they wait. One man is in front of a woman and behind a woman. A woman is on the end of the line. A lady is standing behind the two people in conversation. Work out the order.

Use blue counters to represent the men and red for the women.

a) Make a trial arrangement. ________________________________

b) Read the problem to see if your 1st trial is correct. YES / NO

c) Make further trials and checks until the correct arrangement is found.

____

2. Four children, Barry, Bernice, Bob and Betan have different amounts of money. Bob has $5 more than Bernice. Bernice has double the amount Betan has. Betan has $3 more than Barry. Barry has four $2 coins. How much does each person have?

Use counters/buttons to work out each person's amount.

a) How much does Barry have?____

b) Betan has ____________

c) Betan × 2 = Bernice or Bernice × 2 = Betan? Circle the correct one.

d) Calculate how much Bernice has. ____

e) Now work out Bob's money. ____ How did you do this?____________________________

____

f) Now list the people with the amounts of money ____________ $____,

____________ $____, ____________ $____ and ____________ $____.

Notice here you combined two strategies: Acting it Out to help you Work Backwards.

Solve this one with no help from me.

3. On the beach I counted exactly 39 people. There was 1 more girl than boys and six more men than women. The number of men equalled the number of boys and girls. How many of each group were on the beach?

a) How are you going to solve this?________________________________

____

(b) Show your guesses and checks here.________________________________

____

____

(c) There were ____ women, ____ men, ____ boys and ____ girls.

Creating a Table

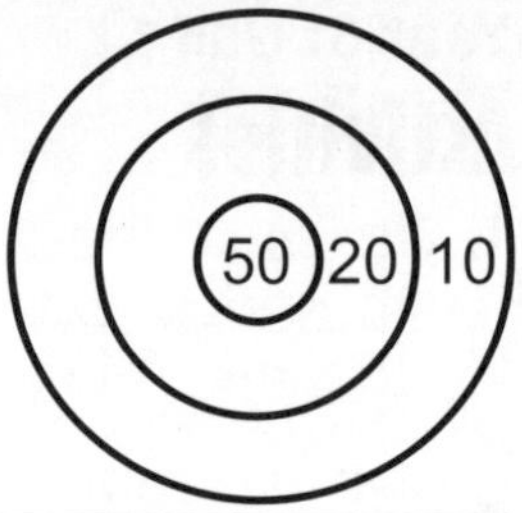

1. Kendo is an archer. He fires three arrows at this target. Create a table to show all of the possible scores he can make. One is done for you.

a)

10	↙↙↙			
20				
50				
Total	30			

b) Is it possible for Kendo to score 130 with 3 arrows? _____

c) Show how the number of combinations would change if he used two rather than three arrows.

10	↙↙	
20		
50		
Total	20	

d) Could he now score 130? YES / NO. How? ______________________________

e) How could Kendo score less than 40 points using 4 arrows? ____________________

____

2. What is (i) the most common total, and (ii) the least common total when 2 six-sided dice are rolled?

Let's get organised and show all of the possible combinations on one grid. I've started it, you finish it off.

	1	2	3	4	5	6
1				5		
2		4				
3						
4					9	
5						
6				10		

a) Write down all of the totals which could be rolled.

____

b) Of these numbers which one appears on the grid the most number of times? ____

c) How many combinations give this number? List them. ____________________

d) Two numbers appear only once on this grid. They are ____ and ____.

e) Why can't one be rolled? ______________________________________

f) Are there four or only two combinations which give a result of 9? ______________

Explain your reasoning. ______________________________________

Activities

1. Stephen and his sister, Andrea, went fishing. They caught bream, whiting and flathead. Of the whiting, they caught double the number of flathead and six more bream than flathead. In all, they caught 54 fish. How many of each variety did they catch?

You could use guessing and checking but creating a table can give you the answer as well.

bream									
whiting									
flathead									
total									

a) What's the least number of flathead they caught? ____

b) If they caught 1 flathead, this means ____ whiting and ____ bream.

c) The total is ____. Enter all of this data on your table.

d) For 2 flathead, bream = ____; whiting = ____; total = ____.

Complete the table. Use patterns to save your working out every cell.

e) Find the column where the total comes to 54. Now record these. ____ bream,

____ whiting, ____ flathead were caught = ____ fish.

f) If they had caught only 12 bream, their total catch would have been ____ whiting

____ flathead (plus the 12 bream) for a total catch of ____.

2. Sixty chairs were arranged in even (same number in each) rows.
What possible ways could these chairs be arranged?

Here you could use counters for chairs and act it out, but try using a table to find the combinations.

Number in row	60											
Number of rows	1											
Total no. of seats	60											

a) Is one row of 60 chairs the same as 60 rows of 1 chair? YES / NO.

b) Will this affect the number of combinations? YES / NO

c) There is a total of ____ possible combinations.

d) Use counters as a model to prove that 6 rows of 12 chairs is not one of the combinations.

____

e) If the chairs were to be placed in a hall with room for only 19 rows, how many

combinations could be used for the hall? ____________

Creating a Table

1. U Beaut chocolate bars cost 59 cents each.

a) Complete the table below showing the cost of these numbers of bars.

No. of U Beaut bars	1	2	3	4	5	6	7	8
Cost	59c		$1.77					

Answer these by using *only* this table - no calculators!

b) How could you find the cost of 12 bars? ______________________________

____

The total cost for 12 is $____.

c) Describe how you'd work out the value of a box of 50 U Beaut bars.

____

A box is worth $____.

d) Deanne said that 15 U Beaut bars would cost $8.75. She calculated this by finding the cost of 7 ($4.13) and the cost of 8 ($4.72) and then added them (7 + 8 = 15) so $4.13 + $4.72 equals the total cost ($8.75). There is an error in her working. Find where she went wrong. ____

e) There were $8.26 worth of U Beaut bars left on the shelf.

How many bars is this equal to? ____

2. A moulding machine can produce 35 plastic toy cars a minute. The materials for these 35 toys costs 72 cents.

(i) How long will it take to produce 315 toy cars?

(ii) If $5.76 is spent on materials, this will produce ____ toy cars.

Let's draw up a table showing time, number of toy cars and the cost.

Time	1	2	3	4	5						
Number	35										
Cost	72 c										

a) Complete the table.

b) Read from your table and find the answers to (i) ____________ and (ii) ____.

Use the table - NO CALCULATORS - for these last three questions.

c) After running for 16 minutes, the machine produced ____ toy cars at a cost of $____.

d) $12.96 was spent in ____ minutes to make ____ toys.

e) 665 toys took ____ minutes at a cost of $____.

Activities

Creating a table is a very logical and methodical way of recording Guessing and Checking to solve a problem.

1. When I add two numbers the total is 46. The product of these two numbers if 493.

The numbers are ____ and ____.

1st no.				
2nd no.				
Sum				
Product				

2. ☐ + ☐ = 41

☐ – ☐ = 5

The numbers are ____ and ____.

1st no.				
2nd no.				
Sum				
Difference				

3. Jane bought 17 stamps, composed of 75 c stamps and 45 c stamps. She spent a total of $10.35 on ____ 75 c stamps and ____ 45 c stamps.

75 c	number			
	value			
45 c	number			
	value			
Total spent		$10.35		

4. On Sale of the Millennium, each contestant starts with $20. $10 is added for a correct answer; $7 is subtracted for a wrong answer. After 25 questions, Jane had $66. How many answers did she give incorrectly?

Starting amount				
No. right				
Winnings				
No. wrong				
Losings				
Balance				

5. Once upon a time, there were three ugly old witches. Esmeralda was twice as old as Agnes, but Agnes was 20 years younger than Sadie. If you added all of their ages together, the total would be 232 years. Find the witches' ages.

Esmeralda				
Agnes				
Sadie				
Total age	232			

6. Set up your own table to help solve this. The sum of two numbers is 121. The difference between the two numbers is 39.

The two numbers are ____ and ____.

Creating a List

Making a list is not random; it must be organised. Check this out.

1. How many three-digit numbers can be created from 4 digits (1, 2, 4 and 6)?

Let's start the list with the first 3 digits - 124.

a) What other 3-digit numbers can begin with 12? ____________________

b) Are there any others that begin with 12? YES / NO

Let's take the 1st and 3rd digit and check these.

c) 14 ☐ and 14 ☐ are the two possibilities.

d) 16 ☐ and 16 ☐ are the next two possibilities.

This isn't the end because we can start with the second digit.

e) 2 ____ ____, 2 ____ ____ are two and 2 ____ ____, 2 ____ ____ are others, as well as 2 ____ ____ and 2 ____ ____.

f) 4 ____ ____, 4 ____ ____, 4 ____ ____, 4 ____ ____, 4 ____ ____ and 4 ____ ____ follow.

g) Write the last 6 combinations. ____________________

Let's record this in a listing.

Complete this.

There is a total of ____ combinations.

124	2 ____ ____	4 ____ ____	
12 ____	2 ____ ____	4 ____ ____	
14 ____	2 ____ ____		
14 ____	2 ____ ____		
16 ____	2 ____ ____		
16 ____	2 ____ ____		

2. The local car dealer is offering any three of these optional extras on new cars at no cost during January. How many different package deals can be created?

Think about this. What is the difference between mag wheels, air, power steering and air, power steering, mag wheels?

JANUARY SALE

Choose any 3 options

FREE

- ☆ Air Conditioning
- ☆ Mag Wheels
- ☆ CD Player
- ☆ Power Steering
- ☆ Glass Tinting

a) Create your list here.

b) There are ____ different package deals.

Activities

1. Jenny has 10 c, 20 c and 50 c coins in her money box. She owes Pauline $1.80. Make a list of all of the possible ways Jenny can give the money to Pauline.

Make sure you use a logical and methodical process so that you don't overlook any of the combinations.

Acting it Out using real coins or counters could help, too.

a) My listing.______________________________

____

____

____

____

____

b) There are ____ different ways.

2. Dad likes a three-course meal at night. Mum spoils him and prepares different meals (by at least one course) from his list of favourites. Make as many different meals as you can.

Favourite Soups
- ☆ Pea & ham
- ☆ Chicken & corn
- ☆ Tomato

Favourite Main Meals
- ☆ Steak & chips
- ☆ Baked Ham & veges
- ☆ Chicken & salad
- ☆ Fish & chips

Favourite Desserts
- ☆ Ice cream & jelly
- ☆ Apple Pie & cream
- ☆ Fruit Salad & custard

a) Complete this tree diagram.

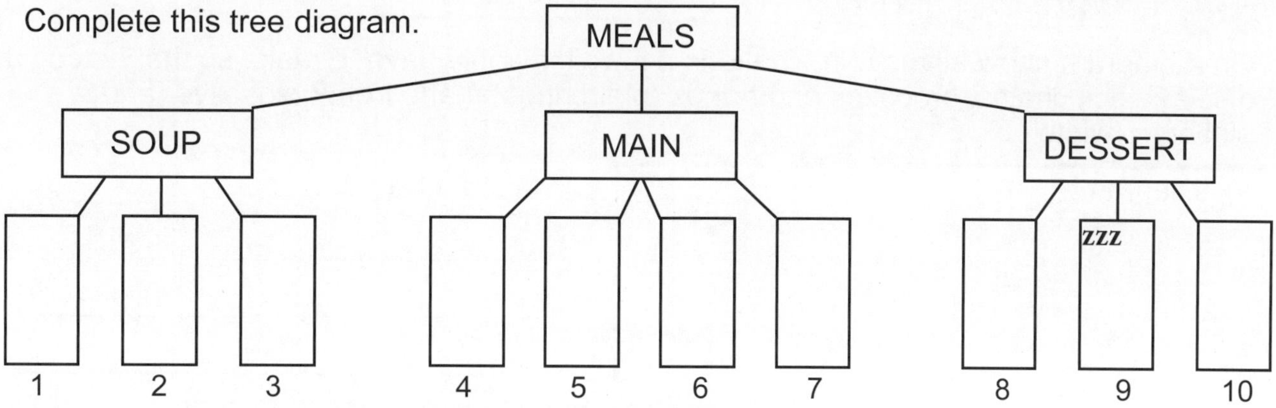

b) Now let's list them in a logical order. Follow my lead.

1, 4, 8______________________________

1, 4, 9______________________________

1, 4, 10______________________________

1, 5, 8______________________________

1, 5, 9______________________________

1, 5, 10______________________________

____

____

____

Dad can have a total of ____ different meals with these combinations.

Year 6:

Test 1

Patterns and Tree Diagrams will help you with these.

1.

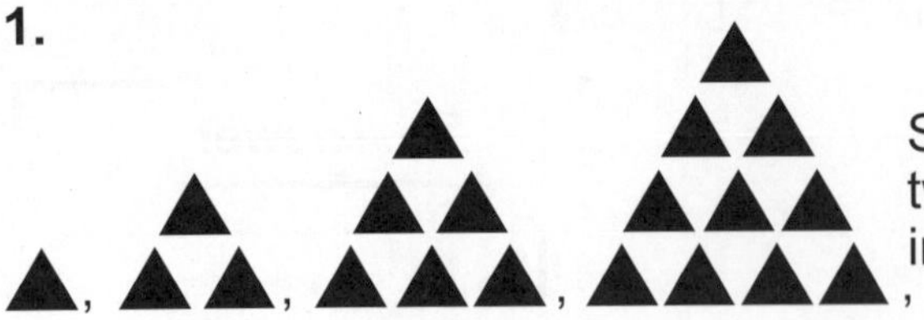

Show the next two elements in this series

a)

and

b)

c) The eighth element would be made up of ____ small triangles in a large __________ shape.

d) Would there be an element in this series that contained 55 small triangles in a large triangular shape? YES / NO

e) Explain why 66 randomly drawn small triangles would *not* be included in this series.

____

2. Use the patterns in these to predict the missing element.

a) 7 ÷ 9 = 0.7777777, 6 ÷ 9 = 0.6666666, 4 ÷ 9 = ______________

b) 33 ÷ 99 = 0.3333333, 34 ÷ 99 = 0.3434343, 35 ÷ 99 = ______________

c) 8 ÷ 11 = 0.7272727, 9 ÷ 11 = 0.8181818, 10 ÷ 11 = ______________

3. On a diagram show all the combinations a jeweller would have to make so that he could display neck chains, bracelets and bands (plain rings) made from gold, silver and platinum (white gold).

a) Diagram

b) Make a list here

____

____

____

____

____

____

____

c) There are ___ combinations.

4. Mrs Packa is very rich. She has many fine rings set with diamonds, rubies, emeralds, opals and sapphires. They are either set in 9 carat gold, 18 carat gold or 22 carat white gold. If Mrs Packa has one of every combination, she must own ____ rings.

Prove this by drawing a tree diagram here to show this number of combinations for the choices given.

Year 6:

Test 2

Try yourself out on these problems. Use either Guessing and Checking or Drawing a Picture to help.

Problem	Working Space	Answer
1. Mr Wright and Master Brian Wright added their ages - total 43 years. The difference in their ages is twenty-one years. Find the ages of Mr Wright and his son, Brian.		
2. During the Olympics, Betta Bus Service has to transport 385 competitors from their accommodation to the hockey venue. Each bus in this company can legally carry no more than 47 passengers. How many buses will be needed? (Use a calculator)		
3. A pool is 12 m long and 7 m wide. It is surrounded by a 3 m paved area on all sides. The whole complex (pool and paving) is enclosed by a 2m high fence. What is the total length of this fencing?		
4. Find the number of marbles in this bag given these clues: (i) there are more than 50 but less than 100. (ii) the number divided by 9 gives an answer with a remainder of 4. (iii) the number ÷ 8 = □ remainder 5.		
5. Don is a cabinetmaker. Mrs Smith gave him 16 seats and 60 legs and asked him to make as many 3- and 4-legged bar stools as he could. How many of each type could Don make?		
6. Gerry's pool was half full of water. After adding another 105 kilolitres the water level had only risen to six tenths full. When this pool is completely full, it will hold a total of _____ kilolitres of water.		

Year 6:

Test 3

Choose from Working Backwards and Acting It Out to help you solve these. Use the spaces for working.

1. I have a big TV box. The length is three times the width. The height is double the width. If I place a metre ruler upright beside the box, the height is 2 cm less than the full ruler. Find the dimensions of this box.

L = ____ cm W = ____m H = ____m

2. There were 9 managers from different companies at the manager business lunch. Each representative exchanged business cards with every other manager. In all, ____ cards were exchanged.

Number of cards =

3. This staircase with 3 steps is made up of 6 blocks. If I had 36 blocks just like these to build a staircase, then there would be ____ steps.

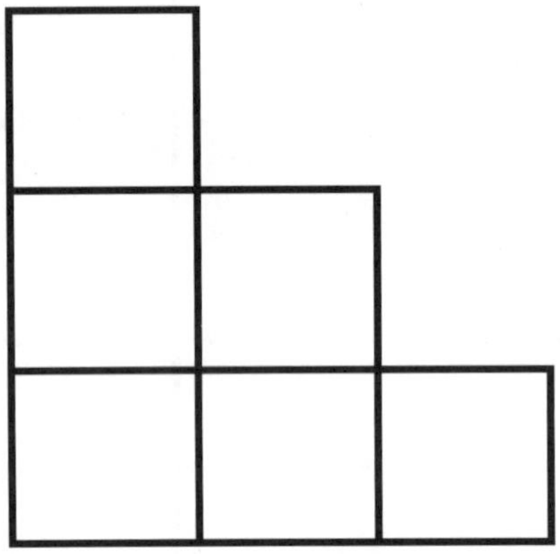

____ steps

4. Charles went to town with $12 and came home with $300. He bought a shirt, some cheese from the delicatessen and visited the dentist that day. He gets paid on a Thursday and cashes his cheque at the bank which is only open on Tuesday, Friday and Saturday. The dentist is not open on Saturday or Sunday. The deli is closed on Thursday and Friday. Work through this information and find out the day Charles went to town.

He went on ____________________.

5. Arrange these coloured balls in order from smallest to largest.

- ☆ Yellow is smaller than red.
- ☆ Orange is bigger than green.
- ☆ Green is not as big as yellow.
- ☆ Red is 3 times the size of green.
- ☆ Green is separated from orange by the two other balls.

Smallest to largest: ____, ____, ____, ____.

6. Jack and Jill cut a big log into two equal pieces with a crosscut saw. It took them 8 minutes of hard work. How long will it take them to cut the log into four equal lengths?

____ minutes

Year 6:

Test 4

Solve these using Making a List or Creating a Table.

1. Gary collects clocks. He has two clocks which keep perfect time but ones goes backwards. Both clocks show the same time at 3 o'clock. When will both again show the same time?

Clock A forwards	3	4					
Clock B backwards	3	2					

They will be the same at ____________.

2. A   K   Q   J   10

a) Here are 5 playing cards. Make a list showing all the possible combinations if 3 cards at a time are chosen.

____

____

____

b) There is a total of ____ combinations.

3. Bicycles and tricycles: 12 seats in all.

a) Complete this table showing all of the combinations possible, then answer the questions.

Bicycles	seats	1											
	wheels	2											
Tricycles	seats	11											
	wheels	33											
Total	seats	12	12	12	12	12	12	12	12	12	12	12	12
	wheels	35											

b) 26 wheels means ____ bikes and ____ trikes.

c) If the number of bikes = trikes then there are ____ wheels.

d) Why is it not possible to have 36 wheels?_______________________________

____

e) How many tricycle wheels are there if there are 8 front wheels on the bicycles?________

Notes